CAMBRIDGE LIBRARY COLLECTION

Books of enduring scholarly value

Darwin

Two hundred years after his birth and 150 years after the publication of 'On the Origin of Species', Charles Darwin and his theories are still the focus of worldwide attention. This series offers not only works by Darwin, but also the writings of his mentors in Cambridge and elsewhere, and a survey of the impassioned scientific, philosophical and theological debates sparked by his 'dangerous idea'.

Foot-Prints of the Creator

The geological writings of Hugh Miller (1802–56) did much to publicise this relatively new science. After an early career in banking in Scotland, Miller became editor of a newly founded Edinburgh newspaper, The Witness, in which he published a series of his own articles based on his geological research, a collection of which was issued as a book, The Old Red Sandstone, in 1841, and led to the Devonian geological period becoming known as the 'Age of the Fishes'. Foot-Prints of the Creator (1849) described his reconstruction of the extinct fish he had discovered in the Old Red Sandstone and argued, on theological grounds, that their perfection of development disproved the current Lamarckian theory of evolution. The book, illustrated with woodcuts, was written partly as a response to the then anonymous Vestiges of the Natural History of Creation, also reissued in this series.

Cambridge University Press has long been a pioneer in the reissuing of out-of-print titles from its own backlist, producing digital reprints of books that are still sought after by scholars and students but could not be reprinted economically using traditional technology. The Cambridge Library Collection extends this activity to a wider range of books which are still of importance to researchers and professionals, either for the source material they contain, or as landmarks in the history of their academic discipline.

Drawing from the world-renowned collections in the Cambridge University Library, and guided by the advice of experts in each subject area, Cambridge University Press is using state-of-the-art scanning machines in its own Printing House to capture the content of each book selected for inclusion. The files are processed to give a consistently clear, crisp image, and the books finished to the high quality standard for which the Press is recognised around the world. The latest print-on-demand technology ensures that the books will remain available indefinitely, and that orders for single or multiple copies can quickly be supplied.

The Cambridge Library Collection will bring back to life books of enduring scholarly value (including out-of-copyright works originally issued by other publishers) across a wide range of disciplines in the humanities and social sciences and in science and technology.

Foot-Prints of the Creator

Or, the Asterolepis of Stromness

HUGH MILLER

CAMBRIDGE UNIVERSITY PRESS

Cambridge, New York, Melbourne, Madrid, Cape Town, Singapore,
São Paolo, Delhi, Dubai, Tokyo

Published in the United States of America by Cambridge University Press, New York

www.cambridge.org
Information on this title: www.cambridge.org/9781108005531

This edition first published 1849
This digitally printed version 2009

ISBN 978-1-108-00553-1 Paperback

FOOT-PRINTS OF THE CREATOR:

OR,

THE ASTEROLEPIS OF STROMNESS.

BY

HUGH MILLER,

AUTHOR OF "THE OLD RED SANDSTONE," ETC.

"When I asked him [Sir Isaac] how this earth could have been re-peopled if ever it had undergone the same fate it was threatened with by the comet of 1680, he answered,—'that required the power of a Creator.'"

(*Mr Conduit's Notes of a Conversation with Sir Isaac Newton.*)

LONDON:

JOHNSTONE AND HUNTER,

26, PATERNOSTER ROW; AND 15, PRINCES STREET, EDINBURGH.

MDCCCXLIX.

TO

SIR PHILIP DE MALPAS GREY EGERTON,

BART. M.P., F.R.S. & G.S.

TO you, Sir, as our highest British authority on fossil fishes, I take the liberty of dedicating this little volume. In tracing the history of Creation, as illustrated in that ichthyic division of the vertebrata which is at once the most ancient and the most extensively preserved, I have introduced a considerable amount of fact and observation, for the general integrity of which my appeal must lie, not to the writings of my friends the geologists, but to the strangely significant record inscribed in the rocks, which it is their highest merit justly to interpret and faithfully to transcribe. The ingenious and popular author whose views on Creation I attempt controverting, virtually carries his appeal from science to the want of it. I would fain adopt an opposite course: And my use, on this occasion, of your name, may serve to evince the desire which I entertain that the collation of my transcripts of hitherto uncopied portions of the geologic history with the history itself, should be in the hands of men qualified, by original vigour of faculty and the patient research of years, either to detect the erroneous or to certify the true. Further, I feel peculiar pleasure in availing myself of the opportunity furnished me by the publication of this little work, of giving expression to my sincere respect for one who, occupying a high place in society, and deriving his descent from names illustrious in history, has wisely taken up the true position of birth and rank in an enlightened country and age; and who, in asserting, by his modest, persevering labours, his proper standing in the scientific world, has rendered himself first among his countrymen in an interesting department of Natural Science, to which there is no aristocratic or "royal road."

I have the honour to be, Sir,

With admiration and respect,

Your obedient humble servant.

HUGH MILLER.

TO THE READER.

THERE are chapters in this little volume which will, I am afraid, be deemed too prolix by the general reader, and which yet the geologist would like less were there any portion of them away. They refer chiefly to organisms not hitherto figured nor described, and must owe their modicum of value to that very minuteness of detail which, by critics of the merely literary type, unacquainted with fossils, and not greatly interested in them, may be regarded as a formidable defect, suited to overlay the general subject of the work. Perhaps the best mode of compromising the matter may be to intimate, as if by beacon, at the outset, the more repulsive chapters; somewhat in the way that the servants of the Humane Society indicate to the skater who frequents in winter the lakes in the neighbourhood of Edinburgh, those parts of the ice on which he might be in danger of losing himself. I would recommend, then, readers not particularly palæontological, to pass but lightly over the whole of my fourth and fifth chapters, with the latter half of the third, marking, however, as they skim the pages, the conclusions at which I arrive regarding the bulk and organization of the extraordinary animal described, and the data on which these are founded. My book, like an Irish landscape dotted with green bogs, has its portions on which it may be perilous for the unpractised surveyor to make any considerable stand, but across which he may safely take his sights and lay down his angles.

It will, I trust, be found, that in dealing with errors which, in at least their primary bearing, affect questions of science, I have not offended against the courtesies of scientific controversy. True, they are errors which also involve moral consequences. There is a species of superstition which inclines men to take on trust whatever assumes the name of science; and which seems to be a re-action on the old superstition, that had faith in witches, but none in Sir Isaac Newton, and believed in ghosts, but failed to credit the Gregorian calendar. And, owing mainly to the wide diffusion of this credulous spirit of the modern type, as little disposed to examine what it receives as its ancient unreasoning predecessor, the development doctrines are doing much harm on both sides of the Atlantic, especially among intelligent mechanics, and a class of young men engaged in the subordinate departments of trade and the law. And the harm, thus considerable in amount, must be necessarily more than merely considerable in degree. For it invariably happens, that when persons in these walks become materialists, they become also turbulent subjects and bad men. That belief in the existence after death, which forms the distinguishing *instinct* of humanity, is too essential a part of man's moral constitution not to be missed when away; and so, when once fairly eradicated, the life and conduct rarely fail to betray its absence. But I have not, from any consideration of the mischief thus effected, written as if arguments, like cannon-balls, could be rendered more formidable than in the cool state by being made red-hot. I have not even felt, in discussing the question, as if I had a man before me as an opponent; for though my work contains numerous references to the author of the "Vestiges," I have invariably thought on these occasions, not of the anonymous writer of the volume, of whom I know nothing, but simply of an ingenious, well-written book, unfortunate in its facts, and not always very happy in its reasonings. Further, I do not think that palæontological fact, in its bearing on the points at issue, is of such a doubtful complexion as to leave the geologist, however much from moral considerations in earnest in the matter, any very serious excuse for losing his temper.

In my reference to the three great divisions of the geologic scale, I designate as *Palæozoic*, all the fossiliferous rocks, from the first appearance of organic existence down to the close of the Permian system; all as *Secondary*, from the close of the Permian system down to the close of the Cretaceous deposits; and all as *Tertiary*, from the close of the Cretaceous deposits down to the introduction of man. The wood-cuts of the volume, of which at least nine-tenths of the whole represent objects never figured before, were drawn and cut by Mr John Adams of Edinburgh (8, Heriot Place), with a degree of care and skill which has left me no reason to regret my distance from the London artists and engravers. So far at least as the objects could be adequately represented on wood, and in the limited space at Mr Adams' command, their truth is such, that I can safely recommend them to the palæontologist. In the accompanying descriptions, and in my statements of geologic fact in general, it will, I hope, be seen that I have not exaggerated the peculiar features on which I have founded, nor rendered truth partial in order to make it serve a purpose. Where I have reasoned and inferred, the reader will of course be able to judge for himself whether the argument be sound or the deduction just; and to weigh, where I have merely speculated, the probability of the speculation; but as, in at least *some* of my statements of fact, he might lie more at my mercy, I have striven in every instance to make these adequately representative of the actualities to which they refer. And so, if it be ultimately found that on some occasions I have misled others, it will, I hope, be also seen to be only in cases in which I have been mistaken myself. The first or popular title of my work, "Foot-prints of the Creator," I owe to Dr Hetherington, the well-known historian of the Church of Scotland. My other various obligations to my friends, literary and scientific, the reader will find acknowledged in the body of the volume, as the occasion occurs of availing myself of either the information communicated, or the organism, recent or extinct, lent me or given.

CONTENTS.

LIST OF WOOD-CUTS.

STROMNESS AND ITS ASTEROLEPIS.

THE LAKE OF STENNIS.

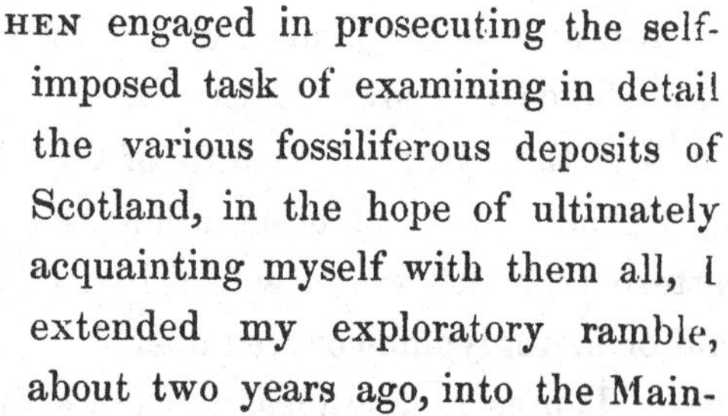

WHEN engaged in prosecuting the self-imposed task of examining in detail the various fossiliferous deposits of Scotland, in the hope of ultimately acquainting myself with them all, I extended my exploratory ramble, about two years ago, into the Mainland of Orkney, and resided for some time in the vicinity of Stromness.

This busy seaport town forms that special centre, in this northern archipelago, from which the structure of the entire group can be most advantageously studied. The geology of the Orkneys, like that of Caithness, owes its chief interest to the immense development which it exhibits of one formation,—the Lower Old Red Sandstone,—and to the extraordinary abundance of its vertebrate remains. It is not too much to affirm, that in the comparatively small portion which this cluster of islands contains of the *third* part of a system regarded only a few years ago as the least fossiliferous in the geologic scale, there are more fossil fish enclosed than in *every* other geologic system in England, Scotland, and Wales, from the Coal Measures to the Chalk inclusive. Orkney is emphatically to the geologist what a juvenile Shetland poetess designates her country, in challenging for it a standing independent of the "Land of Cakes,"—a "Land of Fish;" and, were the trade once fairly opened up, could supply with ichthyolites, by the ton and the shipload, the museums of the world. Its various deposits, with all their strange organisms, have been uptilted from the bottom against a granitic axis, rather more than six miles in length by about a mile in breadth, which forms the great back-bone of the western district of Pomona; and on this granitic axis,—fast jambed in between a steep hill and the sea,—stands the town of Stromness. Situated thus *at the bottom* of the upturned deposits of the island, it occupies exactly such a point of observation as that which the curious eastern traveller would select, in front of some huge pyramid or hieroglyphic-covered obelisk, as a proper site for his tent. It presents, besides, not a few facilities for studying, with the geological phenomena, various interesting points in physical science of a cognate character. Resting on its granitic base, *in front* of the strangely-sculptured pyramid of three broad tiers,—red, black, and gray,—which the Old Red Sandstone of these islands may be regarded as forming, it is but a short half mile from the Great Conglomerate base of the formation, and scarcely a quarter of a mile more from the older beds of its central flagstone deposit; while an hour's sail on the one hand opens to the explorer the overlying arenaceous deposit of Hoy, and an hour's walk on the other introduces him to the Loch of Stennis, with its curiously mixed flora and fauna. But of the Loch of Stennis and its productions more anon.

The day was far spent when I reached Stromness; but as I had a fine bright evening still before me, longer by some three or four degrees of north latitude than the midsummer evenings of the south of Scotland, I set out, hammer in hand, to examine the junction of the granite and the Great Conglomerate, where it has been laid bare by the sea along the low promontory which forms the western boundary of the harbour. The granite here is a ternary of the usual components, somewhat intermediate in grain and colour between the granites of Peterhead and Aberdeen; and the conglomerate consists of materials almost exclusively derived from it,—evidence enough of itself, that when this ancient mechanical deposit was in course of forming, the granite,—exactly such a compound then as it is now,—was one of the surface rocks of the locality, and much exposed to disintegrating influences. This conglomerate base of the Lower Old Red Sandstone of Scotland,—which presents, over an area of many thousand square miles, such an identity of character, that specimens taken from the neighbourhood of Lerwick, in Shetland, or of Gamrie, in Banff, can scarce be distinguished from specimens detached from the hills which rise over the Great Caledonian Valley, or from the cliffs immediately in front of the village of Contin,—seems to have been formed in a vast oceanic basin of primary rock,—a Palæozoic Hudson's or Baffin's Bay,—partially surrounded, mayhap, by primary continents, swept by numerous streams, rapid and headlong, and charged with the broken debris of the inhospitable regions which they drained. The graptolite-bearing grauwacke of Banffshire seems to have been the only fossiliferous rock that occurred throughout the entire extent of this ancient northern basin; and its few organisms now serve to open the sole vista through which the geological explorer to the north of the Grampians can catch a glimpse of an earlier period of existence than that represented by the ichthyolites of the Lower Old Red Sandstone.

Very many ages must have passed ere, amid waves and currents, the water-worn debris which now forms the Great Conglomerate could have accumulated over tracts of sea-bottom from ten to fifteen thousand square miles in area, to its present depth of from one to four hundred feet. At length, however, a thorough change took place ; but we can only doubtfully speculate regarding its nature or cause. The bottom of the Palæozoic basin became greatly less exposed. Some protecting circle of coast had been thrown up around it ; or, what is perhaps more probable, it had sunk to a profounder depth, and the ancient shores and streams had receded, through the depression, to much greater distances. And, in consequence, the deposition of rough sand and rolled pebbles was followed by a deposition of mud. Myriads of fish, of forms the most ancient and obsolete, congregated on its banks or sheltered in its hollows ; generation succeeded generation, millions and tens of millions perished mysteriously by sudden death ; shoals after shoals were annihilated ; but the productive powers of nature were strong, and the waste was kept up. But who among men shall reckon the years or centuries during which these races existed, and this muddy ocean of the remote past spread out to unknown and nameless shores around them? As in those great cities of the desert that lie uninhabited and waste, we can but conjecture their term of existence from the vast extent of their cemeteries. We only know that the dark, finely-grained schists in which they so abundantly occur must have been of comparatively slow formation, and that yet the thickness of the deposit more than equals the height of our loftiest Scottish mountains. It would seem as if a period equal to that in which all human history is comprised might be cut out of a corner of the period represented by the Lower Old Red Sandstone, and be scarce missed when away: for every year during which man has lived upon earth, it is not improbable that the *Pterichthys* and its contemporaries may have lived a century. Their last hour, however, at length came. Over the dark-coloured ichthyolitic schists so immensely developed in Caithness and Orkney, there occurs a pale-tinted, unfossiliferous sandstone, which in the island of Hoy rises into hills of from fourteen to sixteen hundred feet in height; and among the organisms of those newer formations of the Old Red which overlie this deposit, not a species of ichthyolite identical with the species entombed in the lower schists has yet been detected. In the blank interval which the arenaceous deposit represents, tribes and families perished and disappeared, leaving none of their race to succeed them, that other tribes and families might be called into being, and fall into their vacant places in the onward march of creation.

Such, so far as the various hieroglyphics of the pile have yet rendered their meanings to the geologist, is the strange story recorded on the three-barred *pyramid* of Stromness. I traced the formation upwards this evening along the edges of the upturned strata, from where the Great Conglomerate leans against the granite, till where it merges into the ichthyolitic flagstones; and then pursued these from older and lower to newer and higher layers, desirous of ascertaining at what distance over the base of the system its more ancient organisms first appear, and what their character and kind. And, embedded in a grayish-coloured layer of hard flag, somewhat less than a hundred yards over the granite, and about a hundred and sixty feet over the upper stratum of the conglomerate, I found what I sought,—a well-marked bone,—in all probability the oldest vertebrate remain yet discovered in Orkney. What, asks the reader, was the character of this ancient organism of the Palæozoic basin?

As shown by its cancellated texture, palpable to the naked eye, and still more unequivocally by the irregular complexity of fabric which it exhibits under the microscope,—by its speck-like life-points or canaliculi, that remind one of air-bubbles in ice,—its branching channels, like minute veins, through which the blood must once have flown,—and its general ground-work of irregular lines of corpuscular fibre, that wind through the whole like currents in a river studded with islands,—it was as truly osseous in its composition as the solid bones of any of the reptiles of the Secondary, or the quadrupeds of the Tertiary periods. And in form it closely resembled a large roofing-nail. With this bone our more practised palæontologists are but little acquainted, for no remains of the animal to which it belonged have yet been discovered in Britain to the south of the Grampians,* nor, except in the Old Red Sandstone of Russia, has it been detect-

---

* Since the above sentence was written and set in type, I have learned that my ingenious friend Mr Charles Peach of the Customs, Fowey, so well known for his palæontological discoveries, has just found in the Devonian system of Cornwall, fragments of what seem to be dermal plates of *Asterolepis*. It is a somewhat curious circumstance, that the two farthest removed extremities of Great Britain,—Cornwall and Caithness,—should be tipped by fossiliferous deposits of the same ancient system, and that organisms which, when they lived, were contemporary, should be found embedded in the rocks which rise over the British Channel on the one extremity, and overhang the Pentland Frith on the other.

ed anywhere on the Continent. Nor am I aware that, save in the accompanying wood-cut (fig. 1), it has ever been figured. The amateur geologists of Caithness and Orkney have, however, learned to recognise it as the "petrified nail." The length of the entire specimen in this instance was five seventh-eighth inches, the transverse breadth of the head two inches and a quarter, and the thickness of the stem nearly three-tenth parts of an inch. This nail-like bone formed a characteristic portion of the *Asterolepis*,—so far as is yet known, the most gigantic ganoid of the Old Red Sandstone, and, judging from the *place* of this fragment, apparently one of the first.

Fig. 1.

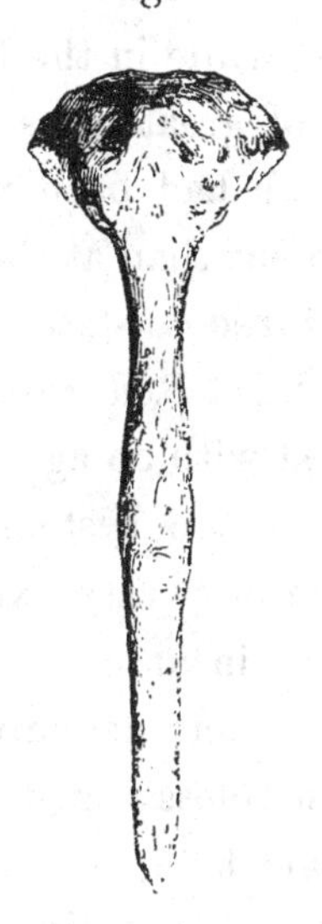

INTERNAL RIDGE OF HYOID PLATE OF ASTEROLEPIS.*

(One-third the natural size, linear.)

---

* Figured from a Thurso specimen, slightly different in its proportions from the Stromness specimen described.

There were various considerations which led me to regard the "petrified nail" in this case as one of the most interesting fossils I had ever seen; and, before quitting Orkney, to pursue my explorations farther to the south, I brought two intelligent geologists of the district,* to mark its place and character, that they might be able to point it out to geological visitors in the future, or, if they preferred removing it to their town museum, to indicate to them the stratum in which it had lain. It showed me, among other things, how unsafe it is for the geologist to base positive conclusions on merely negative data. Founding on the fact that, of many hundred ichthyolites of the Lower Old Red Sandstone which I had disinterred and examined, all were of comparatively small size, while in the Upper Old Red many of the ichthyolites are of great mass and bulk, I had inferred that vertebrate life had been restricted to minuter forms at the commencement than at the close of the system. It had begun, I had ventured to state in the earlier editions of a little work on the "Old Red Sandstone," with an age of dwarfs, and had ended with an age of giants. And now, here, at the very base of the system, unaccompanied by aught to establish the contemporary existence of its dwarfs, —which appear, however, in an overlying bed about a hundred feet higher up,—was there unequivocal proof of the existence of one of the most colossal of its giants. But not unfrequently, in the geologic field, has the practice of basing positive conclusions on merely negative grounds led to a misreading of the record. From evidence of a kind exactly similar to that on which I had built, it was inferred, some two

* Dr George Garson, Stromness, and Mr William Watt, jun., Skaill.

or three years ago, that there had lived no reptiles during the period of the Coal Measures, and no fish in the times of the Lower Silurian System.

I extended my researches, a few days after, in an easterly direction from the town of Stromness, and walked for several miles along the shores of the Loch of Stennis,—a large lake about fourteen miles in circumference, bare and treeless, like all the other lakes and lochs of Orkney, but picturesque of outline, and divided into an upper and lower sheet of water by two low, long promontories, that jut out from opposite sides, and so nearly meet in the middle as to be connected by a thread-like line of road, half-mound, half-bridge. "The Loch of Stennis," says Mr David Vedder, the sailor-poet of Orkney, "is a beautiful Mediterranean in miniature." It gives admission to the sea by a narrow strait, crossed, like that which separates the two promontories in the middle, by a long rustic bridge ; and, in consequence of this peculiarity, the lower division of the lake is salt in its nether reaches and brackish in its upper ones, while the higher division is merely brackish in its nether reaches, and fresh enough in its upper ones to be potable. Viewed from the east, in one of the long, clear, sunshiny evenings of the Orkney summer, it seems not unworthy the eulogium of Vedder. There are moory hills and a few rude cottages in front; and in the background, some eight or ten miles away, the bold, steep mountain masses of Hoy ; while on the promontories of the lake, in the middle distance, conspicuous in the landscape, from the relief furnished by the blue ground of the surrounding waters, stand the tall gray obelisks of Stennis, —one group on the northern promontory, the other on the south,

"Old even beyond tradition's breath."

The shores of both the upper and lower divisions of the lake were strewed, at the time I passed, by a line of *wrack*, consisting, for the first few miles from where the lower loch opens to the sea, of only marine plants, then of marine plants mixed with those of fresh-water growth, and then, in the upper sheet of water, of lacustrine plants exclusively. And the fauna of the loch is, I was informed, of as mixed a character as its flora,—the marine and fresh-water animals having each their own reaches, with certain debateable tracts between, in which each kind expatiates with more or less freedom, according to its specific nature and constitution,—some of the sea-fish advancing far on the fresh water, and others, among the proper denizens of the lake, encroaching far on the salt. The common fresh-water eel strikes out, I was told, farthest into the sea-water ; in which, indeed, reversing the habits of the salmon, it is known in various places to deposit its spawn. It seeks, too, impatient of a low temperature, to escape from the cold of winter, by taking refuge in water brackish enough, in a climate such as ours, to resist the influence of frost. Of the marine fish, on the other hand, I found that the flounder got greatly higher than any of the others, inhabiting reaches of the lake almost entirely fresh. I have had an opportunity elsewhere of observing a curious change which fresh water induces in this fish. In the brackish water of an estuary, the animal becomes, without diminishing in general size, thicker and more fleshy than when in its legitimate habitat the sea : but the flesh loses in quality what it gains in quantity ;—it grows flabby and insipid, and the margin-fin lacks always its strip of transparent fat. But the change induced in the two floras of the lake,—marine and lacustrine,—is considerably more palpable and obvious than that induced in its two faunas. As I passed along the strait, through which it gives admission to the sea, I found the commoner fucoids of our sea-coasts streaming in great luxuriance in the tideway, from the stones and rocks of the bottom. I marked, among the others, the two species of kelp-weed, so well known to our Scotch kelp-burners,—*Fucus nodosus* and *Fucus vesiculosus*,—flourishing in their uncurtailed proportions; and the not inelegant *Halidrys siliquosa*, or "tree in the sea," presenting its amplest spread of pod and frond. A little farther in, *Halidrys* and *Fucus nodosus* disappear, and *Fucus vesiculosus* becomes greatly stunted, and no longer exhibits its characteristic double rows of bladders. But for mile after mile it continues to exist, blent with some of the hardier confervæ, until at length it becomes as dwarfish and nearly as slim of frond as the confervæ themselves; and it is only by tracing it through the intermediate forms that we succeed in convincing ourselves that, in the brown stunted tufts of from one to three inches in length, which continue to fringe the middle reaches of the lake, we have in reality the well-known Fucus before us. Rushes, flags, and aquatic grasses may now be seen standing in diminutive tufts out of the water; and a terrestrial vegetation at least continues to exist, though it can scarce be said to thrive, on banks covered by the tide at full. The lacustrine flora increases, both in extent and luxuriance, as that of the sea diminishes; and in the upper reaches we fail to detect all trace of marine plants: the algæ, so luxuriant of growth along the straits of this "miniature Mediterranean," altogether cease; and a semi-aquatic vegetation attains, in turn, to the state of fullest development anywhere permitted by the temperature of this northern locality. A memoir descriptive of the Loch of Stennis, and its productions, animal and vegetable, such as old Gilbert White of Selborne could have produced, would be at once a very valuable and curious document, important to the naturalist, and not without its use to the geological student.

I know not how it may be with others ; but the special phenomena connected with Orkney that most decidedly bore fruit in my mind, and to which my thoughts have most frequently reverted, were those exhibited in the neighbourhood of Stromness. I would more particularly refer to the characteristic fragment of *Asterolepis,* which I detected in its lower flagstones, and to the curiously-mixed, semi-marine, semi-lacustrine vegetation of the Loch of Stennis. Both seem to bear very directly on that development hypothesis,—fast spreading among an active and ingenious order of minds, both in Britain and America, and which has been long known on the Continent,—that would fain transfer the work of creation from the department of miracle to the province of natural law, and would strike down, in the process of removal, all the old landmarks, ethical and religious.

THE DEVELOPMENT HYPOTHESIS, AND ITS CONSEQUENCES.

Every individual, whatever its species or order, begins and increases until it attains to its state of fullest development, under certain fixed laws, and *in consequence* of their operation. The microscopic monad developes into a fœtus, the fœtus into a child, the child into a man; and, however marvellous the process, in none of its stages is there the slightest mixture of miracle;—from beginning to end, all is progressive development, according to a determinate order of things. Has *Nature*, during the vast geologic periods, been pregnant, in like manner, with the human race? and is the species, like the individual, an effect of progressive development, induced and regulated by law? The assertors of the revived hypothesis of Maillet and Lamarck reply in the affirmative. Nor, be it remarked, is there positive atheism involved in the belief. God might as certainly have *originated* the species by a law of development, as he *maintains* it by a law of development;—the existence of a First Great Cause is as perfectly compatible with the one scheme as with the other: and it may be necessary thus broadly to state the fact, not only in justice to the Lamarckians, but also fairly to warn their non-geological opponents, that in this contest the old anti-atheistic arguments, whether founded on the evidence of design or on the preliminary doctrine of final causes, cannot be brought to bear.

There are, however, beliefs, in no degree less important to the moralist or the Christian than even that in the being of a God, which seem wholly incompatible with the development hypothesis. If, during a period so vast as to be scarce expressible by figures, the creatures now human have been rising, by *almost* infinitesimals, from compound microscopic cells,—minute vital globules within globules, begot by electricity on dead gelatinous matter,—until they have at length become the men and women whom we see around us, we must hold either the monstrous belief, that all the vitalities, whether those of monads or of mites, of fishes or of reptiles, of birds or of beasts, are individually and inherently immortal and undying, or that human souls are *not* so. The difference between the dying and the undying,—between the spirit of the brute that goeth downward, and the spirit of the man that goeth upward,—is not a difference infinitesimally, or even atomically *small.* It possesses all the breadth of the eternity to come, and is an *infinitely great* difference. It cannot, if I may so express myself, be shaded off by infinitesimals or atoms; for it is a difference which—as there can be no class of beings intermediate in their nature between the dying and the undying—admits not of gradation at all. What mind, regulated by the ordinary principles of human belief, can possibly hold that every one of the thousand vital points which swim in a drop of stagnant water are inherently fitted to maintain their individuality throughout eternity? Or how can it be rationally held that a mere progressive step, in itself no greater or more important than that effected by the addition of a single brick to a house in the building state, or of a single atom to a body in the growing state, could ever have produced immortality?

And yet, if the *spirit* of a monad or of a mollusc be not immortal, then must there either have been a point in the history of the species at which a dying brute—differing from its offspring merely by an inferiority of development, represented by a few atoms, mayhap by a single atom—produced an undying man, or man in his present state must be a mere animal, possessed of no immortal soul, and as irresponsible for his actions to the God before whose bar he is, in consequence, never to appear, as his presumed relatives and progenitors the beasts that perish. Nor will it do to attempt escaping from the difficulty, by alleging that God at some certain link in the chain *might* have converted a mortal creature into an immortal existence, by breathing into it a "living soul;" seeing that a renunciation of any such direct interference on the part of Deity in the work of creation forms the prominent and characteristic feature of the scheme,—nay, that it constitutes the very nucleus round which the scheme has originated. And thus, though the development theory be not atheistic, it is at least practically tantamount to atheism. For, if man be a dying creature, restricted in his existence to the present scene of things, what does it really matter to him, for any one moral purpose, whether there be a God or no? If in reality on the same religious level with the dog, wolf, and fox, that are by nature *atheists*,—a nature most properly coupled with irresponsibility,—to what one practical purpose should he know or believe in a God whom he, as certainly as they, is never to meet as his Judge? or why should he square his conduct by the requirements of the moral code, farther than a low and convenient expediency may chance to demand?*

* The Continental assertors of the development hypothesis are greatly more frank than those of our own country regarding the

Nor does the purely Christian objection to the development hypothesis seem less, but even more insuperable than that derived from the province of natural theology. The belief which is perhaps of all others most fundamentally essential to the revealed scheme of salvation, is the belief that "God created man upright," and that man, instead of proceeding onward and upward from this high and fair beginning, to a yet higher and fairer standing in the scale of creation, sank, and became morally lost and degraded. And hence the necessity for that second dispensation of recovery and restoration which forms the entire burden of God's revealed message to man. If, according to the development theory, the

---

"life after death," and what man has to expect from it. The individual, they tell us, perishes for ever; but, then, out of his remains there spring up other vitalities. The immortality of the soul is, it would seem, an idle figment, for there really exist no such things as souls; but is there no comfort in being taught, instead, that we are to resolve into monads and maggots? Job solaced himself with the assurance that, even after worms had destroyed his body, he was in the flesh to see God. Had Professor Oken been one of his comforters, he would have sought to restrict his hopes to the prospect of living in the worms. "If the organic fundamental substance *consist* of infusoria," says the Professor, "so must the whole organic world *originate* from infusoria. Plants and animals can only be metamorphoses of infusoria. This being granted, so also must all organizations *consist* of infusoria, and, during their destruction, dissolve into the same. Every plant, every animal, is converted by maceration into a mucous mass; this putrifies, and the moisture is stocked with infusoria. Putrefaction is nothing else than a division of organisms into infusoria,—a reduction of the higher to the primary life. * * * Death is no annihilation, but only a change. One individual emerges out of another. Death is only a transition to another life,—not into death. This transition from one life to another takes place through the primary condition of the organic, or the mucus."—*Physio-Philosophy*, pp. 187–189.

progress of the "first Adam" was an upward progress; the existence of the "second Adam,"—that "happier man," according to Milton, whose special work it is to "restore" and "regain the blissful seat" of the lapsed race,—is simply a meaningless anomaly. Christianity, if the development theory be true, is exactly what some of the more extreme Moderate divines of the last age used to make it,—an idle and unsightly excrescence on a code of morals that would be perfect were it away.

I may be in error in taking this serious view of the matter: and, if so, would feel grateful to the man who could point out to me that special link in the chain of inference at which, with respect to the bearing of the theory on the two theologies,—natural and revealed,—the mistake has taken place. But if I be in error at all, it is an error into which I find not a few of the first men of the age,—represented, as a class, by our Professor Sedgwicks and Sir David Brewsters, —have also fallen; and until it be shown to *be* an error, and that the development theory is in no degree incompatible with a belief in the immortality of the soul,—in the responsibility of man to God as the final Judge,—or in the Christian scheme of salvation,—it is every honest man's duty to protest against any *ex parte* statement of the question, that would insidiously represent it as ethically an indifferent one, or as unimportant in its theologic bearing, save to "little religious sects and scientific coteries." In an address on the fossil flora, made in September last by a gentleman of Edinburgh, to the St Andrew's Horticultural Society, there occurs the following passage on this subject:—"Life is governed by external conditions, and new conditions imply new races; but then, as to their creation, that is the '*mystery of mysteries.*' Are they created by an immediate fiat and direct act of the

Almighty? or has He originally impressed life with an elasticity and adaptability, so that it shall take upon itself new forms and characters, according to the conditions to which it shall be subjected? Each opinion has had, and still has, its advocates and opponents; but the truth is, that *science*, so far as it knows, or rather so far as it has had the honesty and courage to avow, has yet been unable to pronounce a satisfactory decision. *Either way, it matters little, physically or morally;* either mode implies the same omnipotence, and wisdom, and foresight, and protection; and it is only your little religious sects and scientific coteries which make a pother about the matter,—sects and coteries of which it may be justly said, that they would almost exclude God from the management of his own world, if not managed and directed in the way that they would have it." Now, this is surely a most unfair representation of the consequences, ethical and religious, involved in the development hypothesis. It is not its compatibility with belief in the existence of a First Great Cause that has to be established, in order to prove it harmless; but its compatibility with certain other all-important beliefs, without which simple Theism is of no moral value whatever,—a belief in the immortality and responsibility of man, and in the scheme of salvation by a Mediator and Redeemer. Dissociated from these beliefs, a belief in the existence of a God is of as little *ethical* value as a belief in the existence of the great sea-serpent.

Let us see whether we cannot determine what the testimony of Geology on this question of creation by development really is. It is always perilous to under-estimate the strength of an enemy; and the danger from the development hypothesis to an ingenious order of minds, smitten with the novel fascinations of physical science, has been under-estimated very considerably indeed. Save by a few studious men, who to the cultivation of Geology and the cognate branches add some acquaintance with metaphysical science, the general correspondence of the line of assault taken up by this new school of infidelity, with that occupied by the old, and the consequent ability of the assailants to bring, not only the recently forged, but also the previously-employed artillery into full play along its front, has not only not been marked, but even not so much as suspected. And yet, in order to show that there actually *is* such a correspondence, it can be but necessary to state, that the great antagonist points in the array of the opposite lines are simply the *law* of development *versus* the *miracle* of creation. The evangelistic Churches cannot, in consistency with their character, or with a due regard to the interests of their people, slight or overlook a form of error at once exceedingly plausible and consummately dangerous, and which is telling so widely on society, that one can scarce travel by railway or in a steam-boat, or encounter a group of intelligent mechanics, without finding decided trace of its ravages.

But ere the Churches can be prepared competently to deal with it, or with the other objections of a similar class which the infidelity of an age so largely engaged as the present in physical pursuits will be from time to time originating, they must greatly extend their educational walks into the field of physical science. The mighty change which has taken place during the present century in the direction in which the minds of the first order are operating, though indicated on the face of the country in characters which cannot be mistaken, seems to have too much escaped the notice of our theologians. Speculative theology and the metaphysics are cognate branches of the same science ; and when, as in the last and the preceding ages, the higher philosophy of the world was metaphysical, the Churches took ready cognizance of the fact, and, in due accordance with the requirements of the time, the battle of the Evidences was fought on metaphysical ground. But, judging from the preparations made in their colleges and halls, they do not now seem sufficiently aware, —though the low thunder of every railway, and the snort of every steam-engine, and the whistle of the wind amid the wires of every electric telegraph, serve to publish the fact, —that it is in the departments of physics, not of metaphysics, that the greater minds of the age are engaged,—that the Lockes, Humes, Kants, Berkeleys, Dugald Stewarts, and Thomas Browns, belong to the past,—and that the philosophers of the present time, tall enough to be seen all the world over, are the Humboldts, the Aragos, the Agassizes, the Liebigs, the Owens, the Herschels, the Bucklands, and the Brewsters. In that educational course through which, in this country, candidates for the ministry pass, in preparation for their office, I find every group of great minds which has in turn influenced and directed the mind of Europe for the last three centuries, represented, more or less adequately, save the last. It is an epitome of all kinds of learning, with the exception of the kind most imperatively required, because most in accordance with the genius of the time. The restorers of classic literature,—the Buchanans and Erasmuses,—we see represented in our Universities by the Greek and what are termed the Humanity courses ; the Galileos, Boyles, and Newtons, by the Mathematical and Natural Philosophy courses; and the Lockes, Kants, Humes, and Berkeleys, by the Metaphysical course. But the Cuviers, the Huttons, the Cavendishes, and the Watts, with their successors the practical philosophers of the present age,—men whose achievements in physical science we find marked on the surface of the country in characters which might be read from the moon,—are *not* adequately represented;—it would be perhaps more correct to say, that they are not represented at all;* and the clergy as a class suffer themselves to linger far in the rear of an intelligent and accomplished laity,—a full age behind the requirements of the time. Let them not shut their eyes to the danger which is obviously coming. The battle of the Evidences will have as certainly to be fought on the field of physical science, as it was contested in the last age on that of the metaphysics. And on this new arena the combatants will have to employ new weapons, which it will be the privilege of the challenger to choose. The old, opposed to these, would prove but of little avail. In an age of muskets and artillery, the bows and arrows of an obsolete school of warfare would be found greatly less than sufficient in the field of battle, for purposes either of assault or defence.

"There are two kinds of generation in the world," says Professor Lorenz Oken, in his "Elements of Physio-philosophy;" "the creation proper, and the propagation that is sequent thereupon,—or the *generatio originaria* and *secundaria*. Consequently, no organism has been created of larger size than an infusorial point. No organism is, nor ever has one

---

* I trust that at least by and by there may be an exception claimed, from the general, but, I am sure, well-meant, censure of this passage, in favour of the Free Church of Scotland. It has got as its Professor of Physical Science,—thanks to the sagacity of Chalmers,—Dr John Fleming, a man of European reputation; and all that seems further necessary, in order to secure the benefits contemplated in the appointment, is, that attendance on his course should be rendered imperative on *all* Free Church candidates for the ministry.

been, created, which is not microscopic. Whatever is larger has not been created, but developed. Man has not been created, but developed." Such, in a few brief dogmatic sentences, is the development theory. What, in order to establish its truth, or even to render it in some degree probable, ought to be the geological evidence regarding it? The reply seems obvious. In the first place, the earlier fossils ought to be very *small* in size; in the second, very *low* in organization. In cutting into the stony womb of nature, in order to determine what it contained mayhap millions of ages ago, we must expect, if the development theory be true, to look upon mere embryos and fœtuses. And if we find, instead, the full grown and the mature, then must we hold that the testimony of Geology is not only *not in accordance* with the theory, but in positive opposition to it. Such, palpably, is the *principle* on which, in this matter, we ought to decide. What are the *facts*?

The oldest organism yet discovered in the most ancient geological system of Scotland in which vertebrate remains occur, *seems* to be the *Asterolepis* of Stromness. After the explorations of many years over a wide area, I have detected none other equally low in the system; nor have I ascertained that any brother-explorer in the same field has been more fortunate. It is, up to the present time, the most ancient Scotch witness of the great class of fishes that can in this case be brought into court; nay, it is in all probability the oldest *ganoid* witness the world has yet produced; for there appears no certain trace of this order of fishes in the great Silurian system which lies underneath, and in which, so far as geologists yet know, organic existence first began. How, then, on the two relevant points,—bulk and organization,—does it answer to the demands of the development hypothesis? Was it a mere fœtus of the finny tribe, of minute size, and imperfect, embryotic faculty? Or was it of at least the ordinary bulk, and, for its class, of the average organization? May I solicit the forbearance of the non-geological reader, should my reply to these apparently simple questions seem unnecessarily prolix and elaborate? Peculiar opportunities of observation, and the possession of a set of unique fossils, enable me to submit to our palæontologists a certain amount of information regarding this ancient ganoid, which they will deem at once interesting and new; and the bearing of my statements on the general argument will, I trust, become apparent as I proceed.

THE RECENT HISTORY OF THE ASTEROLEPIS.

ITS FAMILY.

It had been long known to the continental naturalists, that in certain Russian deposits, very extensively developed, there occur in considerable abundance certain animal organisms; but for many years neither their position nor character could be satisfactorily determined. By some they were placed too high in the scale of organized being; by others too low. Kutorga,— a writer not very familiarly known in this country,— described the remains as those of mammals;—the Russian rocks contained, he said, bones of quadrupeds, and, in especial, the teeth of swine: whereas Lamarck, a better known authority, though not invariably a safe one,—for he had a trick of dreaming when wide awake, and of calling his dreams philosophy,—assigned to them a place among the corals. They belonged, he asserted, as shown by certain star-like markings with which they are fretted, to the Polyparia. He even erected for their reception a new genus of Astræ, which he designated, from the little rounded hillock which rises in the middle of each star, the genus *Monticularia*. It was left to a living naturalist, M. Eichwald, to fix their true position zoologically among the class of fishes, and to Sir Roderick Murchison to determine their position geologically as ichthyolites of the Old Red Sandstone.

Sir Roderick, on his return from his great Russian campaigns,—in which he fared far otherwise than Napoleon, and accomplished more,—submitted to Agassiz a series of fragments of these gigantic ganoids ; and the celebrated ichthyologist, who had been introduced little more than a twelvemonth before to the *Pterichthys* of Cromarty, was at first inclined to regard them as the remains of a large cuirassed fish of the Cephalaspian type, but generically new. Under this impression he bestowed upon the yet unknown ichthyolite of which they had formed part, the name *Chelonichthys*, from the resemblance borne by the broken plates to those of the carpace and plastron of some of the Chelonians. At this stage, however, the Russian Old Red yielded a set of greatly finer remains than it had previously furnished ; and of these, casts were transmitted by Professor Asmus, of the University of Dorpat, to the British and London Geological Museums, and to Agassiz. " I knew not at first what to do," says the ichthyologist, " with bones of so singular a conformation that I could refer them to no known type." Detecting, however, on their exterior surfaces the star-like markings which had misled Lamarck, and which he had also detected on the lesser fragments submitted to him by Sir Roderick, he succeeded in identifying both the fragments and bones as remains of the same genus ; and on ascertaining that M. Eichwald had bestowed upon it, from these characteristic sculpturings, the generic name *Asterolepis*, or star-scale, he suffered the name which he himself had originated to drop. Even this second name, however, which the ichthyolite still continues to bear, is in some degree founded in error. Its true scales, as I shall by and by show, were not stelliferous, but fretted by a peculiar style of ornament, consisting of waved anastomosing ridges, breaking atop into angular-shaped dots, scooped out internally like the letter V; and were evidently intermediate in their character between the scales which cover the *Glyptolepis* and those of the *Holoptychius*. And the stellate markings which M. Eichwald graphically describes as minute paps rising out of the middle of star-like wreaths of little leaflets, were restricted to the dermal plates of the head.

Agassiz ultimately succeeded in classing the bones which had at first so puzzled him, into two divisions,—interior and dermal; and the latter he divided yet further, though not without first lodging a precautionary protest, founded on the extreme obscurity of the subject, into cranial and opercular. Of the interior bones he specified two,—a super-scapular bone, —(*supra-scapulaire*),—that bone which in osseous fishes completes the scapular arch or belt, by uniting the scapula to the cranium; and a maxillary or upper jaw-bone. But his world-wide acquaintance with existing fishes could lend him no assistance in determining the places of the dermal bones: they formed the mere fragments of a broken puzzle, of which the key was lost. Even in their detached and irreduceable state, however, he succeeded in basing upon them several shrewd deductions. He inferred, in the first place, that the *Asterolepis* was not, as had been at first supposed, a cuirassed fish, which took its place among the Cephalaspians, but a strongly helmed fish of that Celacanth family to which the *Holoptychius* and *Glyptolepis* belong; in the second, that, like several of its bulkier cogeners, it was in all probability a broad, flat-headed animal; and, in the third, that as its remains are found associated in the Russian beds with numerous detached teeth of large size,—the boar-tusks of Kutorga,—which present internally that peculiar microscopic character on which Professor Owen has erected his Dendrodic or tree-toothed family of fishes,—it would in all likelihood be found that both bones and teeth belonged to the same group. "It appears more than probable," he said, "that one day, by the discovery of a head or an entire jaw, it will be shown that the genera *Dendrodus* and *Asterolepis* form but one" As we proceed, the reader will see how justly the ichthyologist assigned to the *Asterolepis* its place among the Celacanths, and how entirely his two other conjectures regarding it have been confirmed. "I have had in general," he concluded, "but small and mutilated fragments of the creature's bones submitted to me, and of these, even the surface ornaments not well preserved; but I hope the immense materials with which the Old Red Sandstone of Russia has furnished the savans of that country will not be lost to science; and that my labours on this interesting genus, incomplete as they are, will excite more and more the attention of geologists, by showing them how ignorant we are of all the essential facts concerning the history of the first inhabitants of our globe."

I know not what the savans of Russia have been doing for the last few years; but mainly through the labours of an intelligent tradesman of Thurso, Mr Robert Dick,—one of those working men of Scotland of active curiosity and well-developed intellect, that give character and standing to the rest,—I am enabled to justify the classification and confirm the conjectures of Agassiz. Mr Dick, after acquainting himself, in the leisure hours of a laborious profession, with the shells, insects, and plants of the northern locality in which he resides, had set himself to study its geology; and with this view he procured a copy of the little treatise on the Old Red Sandstone to which I have already referred, and which was at that time, as Agassiz's Monograph of the Old Red fishes had not yet appeared, the only work specially devoted to the palæontology of the system, so largely developed in the neighbourhood of Thurso. With perhaps a single exception,—for the Thurso rocks do not yet seem to have yielded a *Pterichthys*,—he succeeded in finding specimens, in a state of better or worse keeping, of all the various ichthyolites which I had described as peculiar to the Lower Old Red Sandstone. He found, however, what I had *not* described,—the remains of apparently a very gigantic ichthyolite; and, communicating with me through the medium of a common friend, he submitted to me, in the first instance, drawings of his new set of fossils; and ultimately, as I could arrive at no satisfactory conclusion from the drawings, he with great liberality made over to me the fossils themselves. Agassiz's Monograph was not yet published; nor had I an opportunity of examining, until about a twelvemonth after, the casts, in the British Museum, of the fossils of Professor Asmus. Besides, all the little information, derived from various sources, which I had acquired respecting the Russian *Chelonichthys*,—for such was its name at the time,—referred it to the cuirassed type, and served but to mislead. I was assured, for instance, that Professor Asmus regarded his set of remains as portions of the plates and paddles of a gigantic *Pterichthys* of from twenty to thirty feet in length. And so, as I had recognised in the Thurso fossils the peculiarities of the *Holoptychian* (Celacanth) family, I at first failed to identify them with the remains of the great Russian fish. All the larger bones sent me by Mr Dick were, I found, cerebral; and the scales associated with these indicated, not a cuirass-protected, but a scale-covered body, and exhibited, in their sculptured and broadly imbricated surfaces, the well-marked Celacanth style of disposition and ornament. But though I could *not* recognise in either bones or scales the remains of one ichthyolite more of the Old Red Sandstone, " that could be regarded as manifesting as peculiar a type among fishes as do the Ichthyosauri and Plesiosauri among reptiles,"* I was engaged at the time in a course of enquiry regarding the cerebral development of the earlier vertebrata, that made me deem them scarce less interesting than if I could. Ere, however, I attempt communicating to the reader the result of my researches, I must introduce him, in order that he may be able to set out with me to the examination of the *Asterolepis* from the same starting-point, to the Celacanth family,—indisputably one of the oldest, and not the least interesting, of its order.

So far as is yet known, all the fish of the earliest fossiliferous system belonged to the placoid or "*broad plated*" order, —a great division of fishes, represented in the existing seas by the Sharks and Rays,—animals that to an internal skeleton of cartilage unite a dermal covering of points, plates, or spines of enamelled bone, and have their gills fixed. The dermal or cuticular bones of this order vary greatly in form, according to the species or family : in some cases they even vary, according to their place, on the same individual. Those button-like tubercles, for instance, with an enamelled thorn, bent like a hook, growing out of the centre of each, which run down the back and tail, and stud the pectorals of the thorn-back (*Raia clavata*), differ very much from the smaller thorns, with star-formed bases, which roughen the other parts of the creature's body ; and the bony points which mottle

* Agassiz's description of the *Pterichthys*, as quoted by Humboldt, in his *Cosmos*.

the back and sides of the sharks are, in most of the known species, considerably more elongated and prickly than the points which cover their fins, belly, and snout. The extreme forms, however, of the shagreen tubercle or plate seem to be those of the upright prickle or spine on the one hand, and of the slant-laid, rhomboidal, scale-shaped plate on the other. The minuter thorns of the ray (fig. 2, *a*) exemplify the extreme of the prickly type; the fins, abdomen, and anterior part of the head of the spotted dog-fish (*Scyllium stellare*), are covered by lozenge-shaped little plates, which glisten with enamel, and are so thickly set that they cover the entire surface of the skin, (fig. 3, *b*)—and these seem equally illustrative of the scale-like form. They are shagreen points passing into osseous scales, without, however, becoming really such; though they approach them so nearly in the shape and disposition of their upper disks, that the true scales, also osseous, of the *Acanthodes sulcatus* (fig. 3, *a*), a ganoid of the Coal Measures, can scarce be distinguished from them, even when microscopically examined. It is only when seen in section that the distinctive difference appears. The true scale of the Acanth, though considerably elevated in the centre, seems to have been planted on the skin; whereas the scale-like shagreen of the dog-fish is elevated over it on an osseous pedicle or footstalk (fig. 5, *a*), as a mushroom is elevated over the sward

Fig. 2.

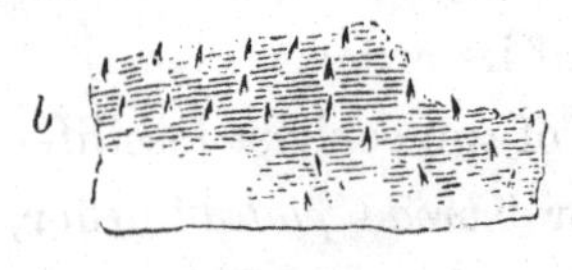

a *Shagreen of the Thornback* (*Raia clavata.*)
b. *Shagreen of Sphagodus,—a placoid of the Upper Silurian.**

* From Murchison's Silurian System.

on its stem; and the base of the stalk is found to resemble in its stellate character that of a shagreen point of the prickly type. The apparent scale is, we find, a bony prickle bent at right angles a little over its base, and flattened into a rhomboidal disk atop.

Fig. 3.

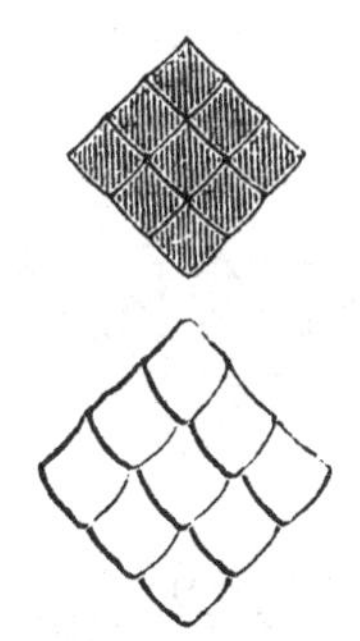

a. *Scales of Acanthodes sulcatus.*
b. *Shagreen of Scyllium stellare (Snout).*
(Mag. eight diameters.)

In small fragments of shagreen (fig. 2, *b*), which have been detected in the bone-bed of the Upper Ludlow Rocks (Upper Silurian), and constitute the most ancient portions of this substance known to the palæontologist, the osseous tubercles are, as in the minuter spikes of the ray, of the upright thorn-like tpye;—they merely serve to show that the placoids of the first period possessed, like those of the existing seas, an ability of secreting solid bone on their cuticular surfaces; and that, though at least such of them as have bequeathed to us specimens of their dermal armature possessed it in the form farthest removed from that of their immediate successors the ganoid fishes, they resembled them not less in the substance of which their dermoskeletal, than in that of which their endoskeletal, parts were composed. For the internal skeleton in both orders, during these early ages, seems to have been equally cartilaginous, and the cuticular skeleton equally osseous. In the ichthyolitic formation immediately over the Silurians,—that of the Lower Old Red Sandstone,—the ganoids first appear; and the members of at least one of the families of the deposit, the Acanths,—a family rich in genera and species,—seem to have formed connecting links between this second order and their placoid predecessors. They were covered with true scales (fig. 4, *a*), and their free gills were protected by gill-covers; and so they must be regarded as real ganoids; but as the shagreen of the spotted dog-fish nearly approaches, in form and character, to ganoid scales, without being really such, the scales of this family, on the other hand, approached equally near, without changing their nature, to the shagreen of the placoids, especially to that of the spiked dog-fish (*Spinax Acanthias.*) (Fig. 4, *b.*) We even find on their under surfaces what seems to be an approximation to the characteristic footstalk. They so considerably thicken in the middle from their edges inwards (fig. 5, *c*), as to terminate in their centres in obtuse points. With these shagreen-like scales, the heads, bodies, and fins of all the species of at least two of the Acanth genera,—*Cheiracanthus* and *Diplacanthus*,—were as thickly covered as the heads, bodies, and fins of the sharks are with their shagreen; and so slight was the degree of imbrication, that the portion of each scale overlaid by the two scales in immediate advance of it did not exceed the one-twelfth part of its entire area. In the scale of the *Cheiracanthus* we find the covered portion indicated by a smooth, narrow band, that ran along its anterior edges, and which the furrows that fretted the exposed surface did

Fig. 4.

a

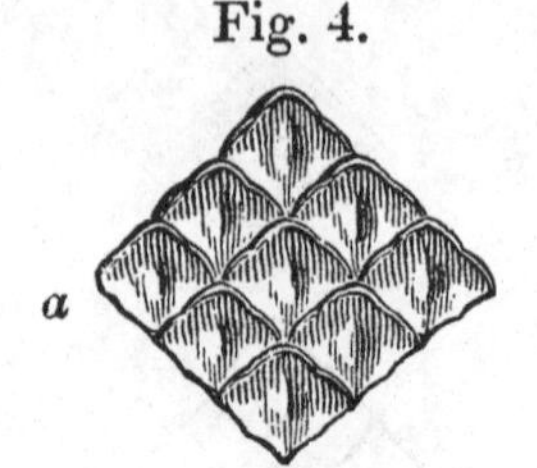

b

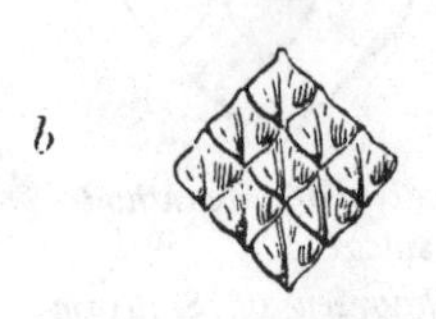

a. *Scales of Cheiracanthus microlepidotus.*
b. *Shagreen of Spinax Acanthias. (Snout.)*
(Mag. eight diameters.)

Fig. 5.

a

b

c

d

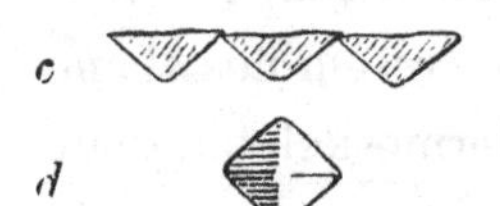

a. *Section of shagreen of Scyllium stellare.*
b. *Under surface of do.*
c. *Section of scales of Cheiracanthus microlepidotus.*
d. *Under surface of do.*
(Mag. eight diameters.)

not traverse. It may be added, that both genera had the anterior edge of their fins armed with strong spines,—a characteristic of several of the Placoid families.

In the Dipterian genera *Osteolepis* and *Diplopterus* the scales were more unequivocally such than in the Acanths, and more removed from shagreen. The under surface of each was traversed longitudinally by a raised bar, which attached it to the skin, and which, in the transverse section, serves to remind one of the shagreen footstalk. They are, besides, of a rhomboidal form; and, when seen in the finer specimens, lying in their proper places on what had been once the creature's body, they seem merely laid down side by side in line, like those rows of glazed tiles that pave a cathedral floor; but on more careful examination, we find that each little tile was deeply grooved on its higher side and end (for it lay diagonally in relation to the head), like the flags of a stone roof (fig. 6, *a*),—that its

Fig. 6.

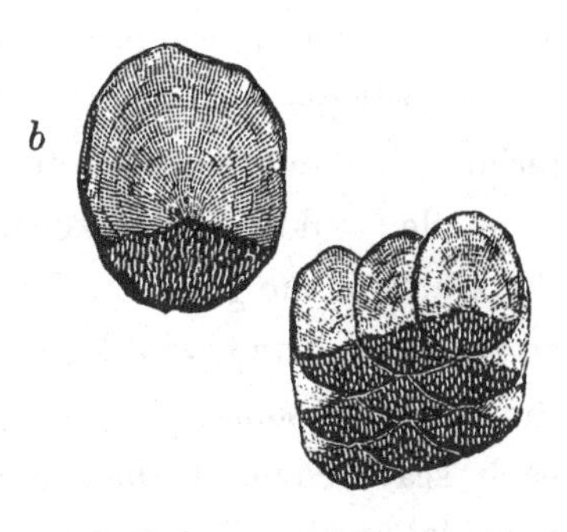

a. *Scales of Osteolepis Microlepidotus.*
b. *Scales of an undescribed species of Glyptolepis.**
(The single scales mag. two diameters;—the others nat. size.)

* These scales, which occur in a detached state, in a stratified clay of the Old Red Sandstone, near Cromarty, present for their size a larger extent of *cover* than the scales of any other ganoid.

lateral and anterior neighbours impinged upon it along these grooves, to the extent of about one-third its area,—and that it impinged, in turn, to the same extent on the scales that bordered on it posteriorly and latero-posteriorly. Now, in the Celacanth family (and on this special point the foregoing remarks are intended to bear), the scales, which were generally of a round or irregularly oval form (fig. 6, *b*), overlapped each other to as great an extent as in any of the existing fishes of the cycloid or ctenoid orders,—to as great an extent, for instance, as in the carp, salmon, or herring. In a slated roof there is no part on which the slates do not lie double, and along the lower edge of each tier they lie triple;—there is more of slate covered than of slate seen: whereas in a tile-roof, the covered portion is restricted to a small strip running along the top and one of the edges of each tile, and the tiles do not lie double in more than the same degree in which the slates lie triple. The scaly cover of the two genera of Dipterians to which I have referred was a cover on the *tile*-roof principle; and this is an exceedingly common characteristic of the scales of the ganoids. The scaly cover of the Celacanths, on the other hand, was a cover on the *slate*-roof principle;—there was in some of their genera about one-third more of each scale covered than exposed; and this is so rare a ganoidal mode of arrangement, that, with the exception of the *Dipterus*,—a genus which, though it gives its name to the Dipterian sept, differed greatly from every other Dipterian,—I know not, beyond the limits of the ancient Celacanth family, a single ganoid that possessed it. The bony covering of the Celacanths was *farthest* removed in character from shagreen, as that of their contemporaries the Acanths approximated to it most nearly; they were, in this respect, the two extremes of their order; and, did we find the

Celacanths in but the later geological formations, while the Acanths were restricted to the earlier, it might be argued by assertors of the development hypothesis, that the amply imbricated, slate-like scale of the latter had been developed in the lapse of ages from the shagreen tubercle, by passing in its downward course,—broadening and expanding as it descended,—through the minute, scarcely imbricated disks of the Acanths, and the more amply imbricated tile-like rhombs of the Dipterians and Palæonisci, until it had reached its full extent of imbrication in the familiar modern type exemplified in both the Celacanths and the ordinary fishes. But such is not the order which nature has observed ;—the two extremes of the ganoid scale appear together in the same early formation ; both become extinct at a period geologically remote ; and the ganoid scales of the existing state of things which most nearly resemble those of ancient time are scales formed on the intermediate or tile-roof principle.

The scales of the Celacanths were, in almost all the genera which compose the family, of great size,—in some species, of the greatest size to which this kind of integument ever attained. Of a Celacanth of the Coal Measures, the *Holoptychius Hibberti*, the scales in the larger specimens were occasionally from five to six inches in diameter. Even in the *Holoptychius Nobilissimus*, in an individual scarcely exceeding two and a half feet in length, they measured from an inch and a half to an inch and three quarters each way. In the splendid specimen of this last species, in the British Museum, there occur but fourteen scales between the ventrals, though these lie low on the creature's body, and the head ; and in a specimen of a smaller species,—the *Holoptychius Andersoni*,—but about seventeen. The exposed portion of the scale was in most species of the family curiously fretted by intermingled ridges and furrows, pits and tubercles, which were either boldly relieved, as in the *Holoptychius*, or existed, as in the *Glyptolepis*, as slim, delicately chiselled threads, lines, and dots. The head was covered by strong plates, which were roughened with tubercles either confluent or detached, or hollowed, as in the *Bothriolepis*, into shallow pits. The jaws were thickly set with an outer range of true fish teeth, and more thinly with an inner range of what seem *reptile* teeth, that stood up, tall and bulky, behind the others, like officers on horseback seen over the heads of their foot-soldiers in front. The *double* fins,—pectorals and ventrals,—were characterized each by a thick, angular, scale-covered centre, fringed by the rays ; and they must have borne externally somewhat the form of the sweeping paddles of the Ichthyosaurian genus,—a peculiarity shared also by the double fins of the *Dipterus*. The *single* fins, in all the members of the family of which specimens have been found sufficiently entire to indicate the fact, were four in number,—an anal, a caudal, and two dorsal fins ; and, with the exception of the anterior dorsal, which was comparatively small, and bent downwards along the back, as if its rays had been distorted when young,* they were all of large size. They crowded thickly on the posterior portion of the body,—the anterior dorsal opposite the ventrals, and the posterior dorsal opposite the anal fin. The fin-rays of the various members of the family, and such of their spinous processes as have been detected, were hollow tubular bones ; or rather, like the larger pieces in the frame-work of the placoids, they were cartilaginous within, and covered externally by a thin osseous crust or shell, which

* A peculiarity which also occurs in the anterior dorsal of the *Dipterus*.

alone survives; and to this peculiarity they owe their family name, Celacanth, or "hollow-spine." The internal hollow, *i.e.* cartilaginous centre, was, however, equally a characteristic of the spinous processes of the *Coccosteus.* In their general proportions, the Celacanths, if we perhaps except one species,—the *Glyptolepis Microlepidotus,*—were all squat, robust, strongly-built fishes, of the Dirk Hatterick or Balfour-of-Burley type; and not only in the larger specimens gigantic in their proportions, but remarkable for the strength and weight of their armour, even when of but moderate stature. The specimen of *Holoptychius Nobilissimus* in the British Museum could have measured little more than three feet from snout to tail when most entire; but it must have been nearly a foot in breadth, and a bullet would have rebounded flattened from its scales. And such was that ancient Celacanth family, of which the oldest of our Scotch ganoids,—the *Asterolepis* of Stromness,—formed one of the members, and which for untold ages has had no living representative.

Let us now enter on our proposed inquiry regarding the cerebral development of the earlier vertebrata, and see whether we cannot ascertain after what manner the first true brains were lodged, and what those modifications were which their protecting box, the cranium, received in the subsequent periods. Independently of its own special interest, the inquiry will be found to have a direct bearing on our general subject.

CEREBRAL DEVELOPMENT OF THE EARLIER VERTEBRATA.

ITS APPARENT PRINCIPLE.

IT is held by a class of naturalists, some of them of the highest standing, that the skulls of the vertebrata consist, like the columns to which they are attached, of vertebral joints, composed each, in the more typical forms of head, as they are in the trunk, of five parts or elements,—the centrum or body, the two spinous processes which enclose the spinal cord, and the two ribs. These cranial vertebræ, four in number, correspond, it is said, to the four senses that have their seat in the head: there is the nasal vertebra, the centrum of which is the vomer, its spinal processes the nasal and ethmoid bones, and its ribs the *upper* jaws; there is the ocular vertebra, the centrum of which is the anterior portion of the sphenoid bone, its spinal processes the frontals, and its ribs the *under* jaws; there is the lingual vertebra, the centrum of which is the posterior sphenoid bone, its spinal processes the parietals, and its ribs the hyoid and branchial bones,—portions of the skeleton largely developed in fishes; and, lastly, there is the auditory vertebra, the centrum of which is the base of the occipital bone, and its spinal processes the occipital crest, and which in the osseous fishes bears attached to it, as its ribs, the bones of the scapular ring. And the cerebral segments thus constructed we find represented in typical diagrams of the skull, as real vertebræ. Professor Owen, in his lately published treatise on "The Nature of Limbs,"—a work charged with valuable fact, and instinct with philosophy,—figures in his draught of the archetypal skeleton of the vertebrata, the four vertebræ of the head, in a form as unequivocally such as any of the vertebræ of the neck or body.

Now, for certain purposes of generalization, I doubt not that the conception may have its value. There are in all nature and in all philosophy certain central ideas of general bearing, round which, at distances less or more remote, the subordinate and particular ideas arrange themselves,

"Cycle and epicycle, orb in orb."

In the classifications of the naturalist, for instance, all *species* range round some central *generic* idea ; all genera round some central idea, to which we give the name of *order* ; all orders round some central idea of *class* ; all classes round some central idea of *division* ; and all divisions round the interior central idea which constitutes a *kingdom*. Sir Joshua Reynolds forms his theory of beauty on this principle of central ideas. "Every species of the animal, as well as of the vegetable creation," he remarks, "may be said to have a fixed or determinate form, towards which nature is continually inclining, like various lines terminating in a centre ; or it may be compared to pendulums vibrating in different directions over one central point, which they all cross, though only one of their number passes through any other point." He instances, in illustrating his theory, the Grecian *beau ideal* of the human nose, as seen in the statues of the Greek deities. It formed a straight line ; whereas all deformity of nose is of a convex or concave character, and occasioned by either a rising above or a sinking below this medial line of beauty. And it may be of use, as it is unquestionably of interest, to conceive, after this manner, of a certain type of skeleton, embodying, as it were, the central or primary type of all vertebral skeletons, and consisting of a double range of rings, united by the bodies of the vertebræ, as the two rings of a figure 8 are united at their point of junction ; the upper ring forming the enclosure of the brain,—spinal, and cephalic ; the lower that of the viscera,—respiratory, circulatory, and digestive. Such is the idea embodied in Professor Owen's archetypal skeleton. It is a series of vertebræ composing double rings,—their *brain*-rings comparatively small in the vertebræ of the trunk, but of much greater size in the vertebræ of the head. But it must not be forgotten, that central ideas, however necessary to the classification of the naturalist, are not historic facts. We may safely hold, with the philosophic painter, that the outline of the typical human nose is a straight line ; but it would be very unsafe to hold, as a consequence, that the first men had all straight noses. And when we find it urged by at least one eminent assertor of the development hypothesis,—Professor Oken,—that light was the main agent in developing the substance of nerve,—that the nerves, ranged in pairs, in turn developed the vertebræ, each vertebra being but "the periphery or envelope of a pair of nerves,"—and that the nerves of those four senses of smell, sight, taste, and hearing, which, according to the Professor, "make up the head," originated the four cranial vertebræ which constitute the skull, —it becomes us to test the central idea, thus converted into a sort of historic myth, by the realities of actual history.

What, then, let us enquire, is the real history of the cerebral development of the vertebrata, as recorded in the rocks of the earlier geologic periods ?

Though the vertebrata existed in the ichthyic form throughout the vastly extended Silurian period, we find in that system no remains of the cranium : the Silurian fishes *seem*, as has been already said (page 29), to have been exclusively placoid ; and the purely cartilaginous box formed by nature for the protection of the brain in this order has in no case been preserved. Teeth, and, in at least one or two instances, the minute jaws over which they were planted have been found, but no portion of the skull. We know, however, that in the fishes of the same order which now exist, the cranium consists of one undivided piece of a cartilaginous substance, set thickly over its outer surface with minute polygonal points of bone (fig. 7), composed internally of star-like rays, that radiate from the centre of ossification, and that present, in consequence, seen through a microscope, the appearance of the polygonal cells of a coral of the genus Astræ. The pattern induced is that of stars set within polygons. Along the sides or top of this unbroken cranial box, that exhibits no mark of suture, we find the perforations through which the nerves of smell, sight, taste, and hearing passed from the brain outwards, and see that they have failed to originate distinct vertebral envelopes for themselves ;—they all lodge in one undivided mansion-house, and have merely separate doors. We find, further,

Fig. 7.

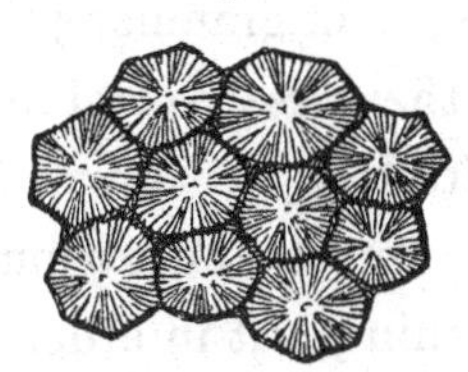

*Osseous points of placoid cranium.**
(Mag. twelve diameters.)

* From the head of *Raia clavata*.

that the homotypal *ribs* of the entire cranium consist, not of four, but simply of a single pair, attached to the occiput, and which serves both to suspend the jaws, upper and nether, in their place under the middle of the head, and to lend support to the hyoid and branchial framework ; while the scapular ring we find existing, as in the higher vertebrata, not as a cerebral, but as a cervical or dorsal appendage. In the wide range of the animal kingdom there are scarce any two pieces of organization that less resemble one another in form than the vertebræ of the placoids resemble their skulls ; and the difference is not merely external, but extends to even their internal construction. In both skull and vertebræ we detect an union of bone and cartilage ; but the bone of each vertebra forms an internal continuous nucleus, round which the cartilage is arranged ; whereas in the skulls it is the cartilage that is internal, and the bone is spread in granular points over it. If we dip the body of one of the dorsal vertebræ of a herring into melted wax, and then withdraw it, we will find it to represent in its crusted state the vertebral centrum of a placoid,—soft without, and osseous within ; but in order to represent the placoid skull, we would have first to mould it out of one unbroken piece of wax, and then to cover it over with a priming of bone-dust. And such is the effect of this arrangement, that, while the skull of a placoid, exposed to a red heat, falls into dust, from the circumstance that the supporting framework on which the granular bone was arranged perishes in the fire, the vertebral centrum, whose internal framework is itself bone, and so *not* perishable, comes out in a state of beautiful entireness,—resembling in the thornback a squat sandglass, elegantly fenced round by the lateral pillars (fig. 8, *b*); and in the dog-fish (*a*) a more elongated sand-glass, in which the lateral pillars are wanting. Such are the heads and vertebral joints of the existing placoids; and such, reasoning from analogy, seem to have been the character and construction of the heads and vertebral joints of the placoids of the Silurian period,—earliest-born of the vertebrata.

Fig. 8.

a b

a. *Osseous centrum of Spinax Acanthias.*
b. *Osseous centrum of Raia clavata.*

(Nat. size.)

The most ancient brain-bearing craniums that have come down to us in the fossil state, are those of the ganoids of the Lower Old Red Sandstone; and in these fishes the true skull appears to have been as entirely a simple cartilaginous box, as that of the placoids of either the Silurian period or of the present time, or of those existing ganoids, the sturgeons. In the Lower Old Red genera *Cheiracanthus* and *Diplacanthus*, though the heads are frequently preserved as amorphous masses of coloured matter, we detect no trace of internal bone, save perhaps in the gill-covers of the first-named genus, which were fringed by from eighteen to twenty minute osseous rays. The cranium seems to have been covered, as in the shark family, by skin, and the skin by minute shagreen-like scales; and all of the interior cerebral framework which appears underneath exists simply as faint impressions of an undivided body, covered by what seem to be osseous points,—the bony molecules, it is probable, which encrusted the cartilage. The jaws, in the better specimens, are also preserved in the same doubtful style; and this state of keeping is the common one in deposits in which every true bone, however delicate, presents an outline as sharp as when it occupied its place in the living animal. The dermal or skin-skeleton of both genera, which consisted, as has been shown (pages 31, 32) of shagreen-like osseous scales and slender spines, both brilliantly enamelled, is preserved entire ; whereas the interior framework of the head exists as mere point-speckled impressions ; and the inference appears unavoidable, that parts which so invariably differ in their state of keeping *now*, must have essentially differed in their substance originally.

Now, in the *Cheiracanthus* we detect the first faint indications of a peculiar arrangement of the dermal skeleton, in relation to certain parts of the skeleton within, which,—greatly more developed in some of its contemporaries,—led to important results in the general structure of these ganoids, and furnishes the true key to the character of the early ganoid head. In such of the existing placoids as I have had an opportunity of examining, the only portions of the dermal skeleton of bone which conform in their arrangement to portions of the interior skeleton of cartilage, are the teeth, which are always laid on a base of skin right over the jaws : there is also an approximation to arrangement of a corresponding kind, though a distant one, in those hook-armed tubercles of certain species of rays which run along the vertebral column ; but in the shagreen by which the creatures are covered I have been able to detect no such arrangement. Whether it occurs on the fins, the body, or the head, or in the scale form, or in that of the prickle, it manifests the same careless irregularity. And on the head and body of the *Cheiracanthus*, and on all its fins save one, the shagreen-like scales, though laid down more symmetrically in lines than true shagreen, manifested an equal absence of arrangement in relation to the frame-work within. On that one fin, however,—the caudal,—the scales, passing from their ordinary rhomboidal to a more rectangular form, ranged themselves in right lines over the internal rays (fig. 9, *a*), and imparted to these such strength as a splint of wood or whalebone fastened over a fractured toe or finger imparts to the injured digit,—a provision which was probably rendered necessary in the case of this important organ of motion, from the circumstance that it was the only fin which the creature possessed that was not strengthened and protected anteriorly by a strong spine. In the *Cheirolepis*,—a contemporary fish, characterized, like its cogeners the *Cheiracanthus* and *Diplacanthus*, by shagreen-like scales, but in which the spines were wanting,—we find a farther development of the provision. In all the fins the richly-enamelled dermal covering was arranged in lines over the rays (fig. 9, *b*) ; and the scale, which assumes in the fins, like the scales on the tail of the *Cheiracanthus*, though somewhat more irregularly, a rectangular shape, is so considerably elongated, that it assumes for its normal character as a scale, that of the joint of an external ray. A similar arrangement of external protection takes place in this genus over the bones of the head : the cartilaginous jaws receive their osseous dermal covering, and, with these, the hyoid bones, the opercules, and the cranium. And it is in these dermal plates, which covered an interior skull, of which, save in one genus,—the *Dipterus*,—not a vestige remains in any of the Old Red fishes thus protected, that we first trace what

Fig. 9.

a. *Portion of caudal fin of Cheiracanthus.**
b. *Portion of caudal fin of Cheirolepis Cummingiæ.*
(Mag. three diameters.)

* The darker, upper patch in this figure indicates a portion in which the scales of the fins in the fossil still retain their enamel ; —the lighter, a portion from which the enamel has disappeared.

seem to be the homologues of the cranial bones of the osseous fishes,—at least their homologues so far as the *cuticular* can represent the *internal*. They appear for the first time, not as modified spinous processes, broadened, as in the carpace of the chelonians, into *osseous* plates, but like those *corneous* external plates of this order of reptiles (known in one species as the tortoise-shell of commerce), the origin of which is purely cuticular, and which evince so little correspondence in their divisions with the sutures of the bones on which they rest, that they have been instanced, in their relation to the joinings beneath, as admirable illustrations of the *cross-banding* of the mechanician.

In the heads of the osseous fishes, the cranium proper, though consisting, like the skulls of birds, reptiles, and mammals, of several bones, exists from snout to nape, and from mastoid to mastoid, as one unbroken box; whereas all the other bones of the head, such as the maxillaries and intermaxillaries, the lower jaws, the opercular appendages, the branchial arches, and the branchiostegous rays, are connected but by muscle and ligament, and fall apart under the putrefactive influences, or in the process of boiling. This unbroken box, which consists, in the cod, of twenty-five bones, is the *homologue* of that cranial box of the placoids which consists of one entire piece, and the *homotype*, according to Oken, of the bodies and spinal processes of four vertebræ; while the looser bones which drop away represent their *ribs*. The upper surface of the box,—that extending from the nasal bone to the nape,—is the only part over which a dermal buckler could be laid, as it is the only part with which the external skin comes in contact; and so it is between this upper surface and the cranial bucklers of the earlier ganoids that we have to institute comparisons. For it is a curious fact, that, with the exception of the Old Red genera *Acanthodus*, *Cheiracanthus*, and *Diplacanthus*,* all the ganoids of the period in which ganoids first appear *have* dermal bucklers placed right over their true skulls, and that these, though as united in their parts as the bones proper to the cranium in quadrupeds and fishes, are composed of several pieces, furnished each with its independent centre of ossification. The Dipterians, the Celacanths, the Cephalaspians, and at least one genus placed rather doubtfully among the Acanths,—the genus *Cheirolepis*, —all possessed cranial bucklers extending from the nape to the snout, in which the plates, various, in the several genera, in form and position, were fast *soldered* together, though in every instance the lines of suture were distinctly marked.

On each side of this external cranium the various cerebral plates, like the corresponding cerebral *ribs* in the osseous fishes, were free, at least not anchylosed together; and some of their number unequivocally performed, in part at least, the functions of two of these cerebral ribs, viz. the upper and under jaws, with the functions of the opercular appendages attached to the latter. In the cod, as in most other osseous fishes, the upper portion of the cranium consists of thirteen bones, which represent, however, only seven bones in the human skull,—the nasal, the frontal, the two parietal, the occipital, and one-half the two temporal bones. And whereas in man, and in most of the mammals, there are four of these placed in the medial line,—the four which, according to the assertors of the vertebral theory, form the spinal crests of the four cerebral vertebræ,—in the cod there are but

* The Acanths of the Coal Measures possess the cranial buckler.

three. The super-occipital bone, A, (fig. 10) pieces on to the superior frontal, C, C, C ; and the parietals, B, B, which in

Fig. 10.

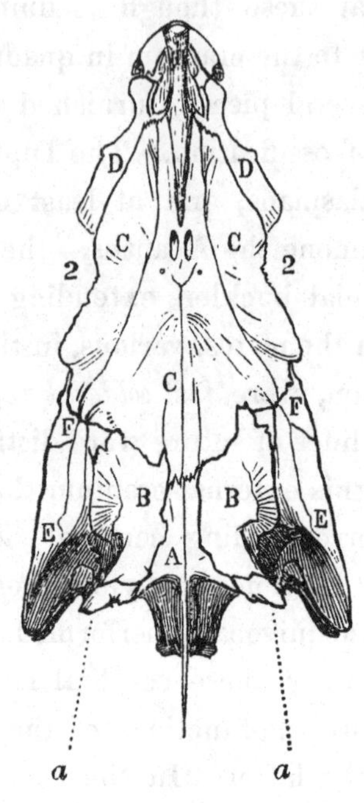

UPPER SURFACE OF CRANIUM OF COD.*

A, *Occipital bone.*	F, F, *Posterior frontals.*
B, B, *Parietals.*	E, E, *Mastoid bones.*
C, C, C, *Superior frontal.*	2, 2, *Eye orbits.*
I, *Nasal bone.*	a, a, *Par-occipital bones.*
D, D, *Anterior frontal.*	

---

* Professor Owen, in fixing the homologies of the ichthyic head, differs considerably from Cuvier ; but his view seems to be demonstrably the correct one. It will, however, be seen, that in my attempted comparison of the divisions of the ancient ganoid cranium with those of the craniums of existing fishes, the points at issue between the two great naturalists are not involved, otherwise than as mere questions of words. The matter to be deter-

the human subject form the upper and middle portions of the cranial vault, are thrust out laterally and posteriorly, and take their places, in a subordinate capacity, on each side of the super-occipital. This is not an invariable arrangement among fishes ;—in the carp genus, for instance, the parietals assume their proper medial place between the occipital and frontal bones ; but so very general is the displacement, that Professor Owen regards it as characteristic of the great ichthyic class, and as the first example in the vertebrata, reckoning from the lower forms upwards, of a sort of natural dislocation among the bones,—" a modification," he remarks, " which, sometimes accompanied by great change of place, has tended most to obscure the essential nature of parts, and their true relations to the archetype."

Of all the cerebral bucklers of the first ganoid period, that which best bears comparison with the cranial front of the cod is the buckler of the *Coccosteus* (fig. 11.) The general proportions of this portion of the ancient Cephalaspian head differ very considerably from those of the corresponding part in the modern cycloid one; but in their larger divisions, the modern and the ancient answer bone to bone. Three osseous plates in the *Coccosteus*, A, C, 1, the homologues, apparently, of the occipital, frontal, and nasal bones, range along the medial line. The apparent homologues of

---

mined, for instance, is not whether plate A in the skulls of the cod and *Coccosteus* be the homologue of a part of the occipital or that of a part of the parietal bones, but whether plate A in the *Coccosteus* be the homologue of plate A in the cod. The letters employed I have borrowed from Agassiz's restoration of the *Coccosteus;* whereas the figures intimate divisions which the imperfect keeping of the specimens on which the ichthyologist founded did not enable him to detect.

Fig. 11.

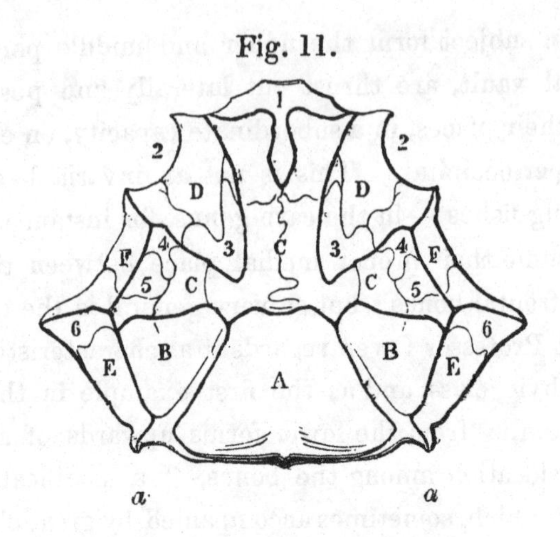

CRANIAL BUCKLER OF COCCOSTEUS DECIPIENS.

a, a, *Points of attachment to the cuirass which covered the upper part of the creature's body.*

the parietals, B, B, occupy the same position of lateral displacement as the parietals of the cod and of so many other fishes. The posterior frontals, F, and the anterior frontals, D, also occupy places relatively the same, though the latter, which are of greater proportional size, encroach much further, laterally and posteriorly, on the superior frontal C, C, C, and sweep entirely round the upper half of the eye-orbits, 2, 2. The apparent homologue of the mastoid bone, E, which also occupies its proper place, joins posteriorly to a little plate, *a*, imperfectly separated in most specimens from the parietal, but which seems to represent the par-occipital bone ; and it is a curious circumstance, that as, in many of the osseous fishes, it is to these bones that the forks of the scapular arch are attached, they unite in the *Coccosteus* in furnishing, in like manner, a point of attachment to the cuirass which covered the upper part of the creature's body. Of the true internal skull of the *Coccosteus* there remains not a vestige. Like that of the sturgeon, it must have been a perishable, cartilaginous box.

In the *Osteolepis*,—an animal the whole of whose external head I have, at an expense of some labour, and from the examination of many specimens, been enabled to restore,—the cranial buckler (fig. 12) was divided in a more arbitrary

Fig. 12.

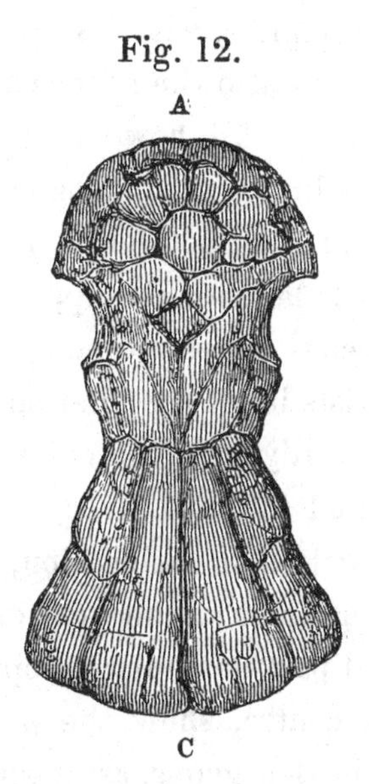

CRANIAL BUCKLER OF OSTEOLEPIS.

style; and we find that an element of uncertainty mingles with our inferences regarding it, from the circumstance that some of its lines of division, especially in the frontal half, were not real sutures, but formed merely a kind of surface-tatooing, resorted to as if for purposes of ornament. The cranial buckler of the *Asterolepis* exhibited, as I shall afterwards have occasion to show, a similar peculiarity;—both had their pseudo-sutures, resembling those false joints introduced by the architect into his rusticated basements, in order to impart the necessary aspect of regularity to what is technically termed the coursing and banding of the fabric. We can, however, determine, notwithstanding the induced obscurity, that the buckler of the *Osteolepis* was divided transversely in the middle into two main parts or segments,—an occipital part, A, and a frontal part, C; and that the occipital segment *seems* to include also the parietal and mastoid plates, and the frontal segment to comprise, with its own proper plates, not only the nasal plate, but also the representative of the anterior part of the vomer. All, however, is obscure. But in our uncertainty regarding the homologies of the divisions of this dermal buckler, let us not forget the homology of the buckler itself, as a whole, with the upper surface of the true cranium in the osseous fishes. Though frequently crushed and broken, it exists in all the finer specimens of my collection as a symmetrically-arranged collocation of enamelled plates, as firmly united into one piece, though they all indicate their distinct centres of ossification, as the corresponding surface of the cranium in the carp or cod. The lateral curves in the frontal part immediately opposite the lozenge-shaped plate in the centre, show the position of the eyes, which were placed in this genus, as in some of the carnivorous turtles, immediately over the mouth,—an arrangement common to almost all the ganoids of the Lower Old Red Sandstone. The nearly semicircular termination of the buckler formed the creature's snout; and in the *Osteolepis*, as in the *Glyptolepis* and the *Diplopterus*, it was armed on the under side, like the vomer of so many of the osseous fishes, with sharp teeth. Some of my specimens indicate the nasal openings a little in advance of the eyes. The nape of the creature was covered by three detached plates (9, 9, 9, fig. 13), which rested upon the anterior dorsal scales, and whose homologues, in the osseous fishes, may possibly be

Fig. 13.

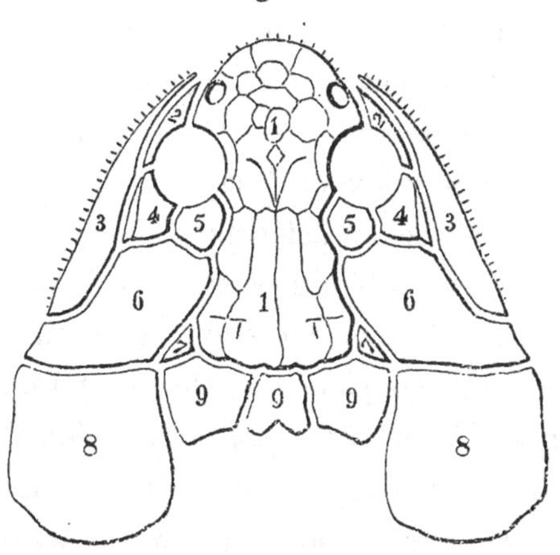

UPPER PART OF HEAD OF OSTEOLEPIS.

found in those bones which, uniting the shoulder-bones to the head, complete the scapular belt or ring. The operculum we find represented by a single plate (8), which had attached to it, as its sub-operculum, a plate (13) of nearly equal size (see figs. 14 and 15). Four small plates (2, 4, 5), formed the under curve of the eyes, described in many of the osseous fishes by a chain of small bones or ossicles; a considerably larger plate (6) occupied the place of the preopercular bone; while the intermaxillaries had their representatives in well-marked plates (3, 3), which, in the genera *Osteolepis*, *Diplopterus*, and *Glyptolepis*, we find bristling so thickly with teeth along their lower edges, as to remind us of the miniature saws employed by the joiner in cutting out circular holes. These external intermaxillaries did not, as in the perch or cod, meet in front of the nasal bone and vomer, but joined on at the side, a little in advance of the eyes, leaving the rounded termination of the cranial buckler, which, like the intermaxillaries, was thickly fringed with teeth, to form, as has been already said, the creature's snout.

The under jaws (10),—strongly-marked bones in at least all the Dipterian and Celacanth genera,—we find represented externally by massy plates, bearing, like those of the upper jaw, their range of teeth. As shown in a well-preserved specimen of the lower jaw of *Holoptychius*, in my possession, they were boxes of bone enclosing a bulky nucleus of cartilage, which, in approaching towards the condyloid process, where great strength was necessary, was thickly traversed by osseous cancilli, and passed at the joint into true bone. It is in the under jaws of the earlier ganoids that we first detect a true union of the external with the internal skeleton,—of the bony plates and teeth, which were *mere plates and teeth of the skin*, with the osseous, granular walls which enclosed at least all the larger pieces of the cartilaginous frame-work of the interior. The jaws of the Rays and Sharks, formed of cartilage, and fenced round on their sides and edges by their thin coverings of polygonal, bony points, are wholly internal and skin-covered; whereas the teeth, which rest on the soft cuticular integument right over them, are as purely dermal as the surrounding shagreen. Teeth and shagreen may, we find, be alike stripped off with the skin. Now, in the earlier ganoidal jaw, two sides of the osseous box which it composed,—its outer and under sides,—were mere dermal plates, representative of the skin of the placoids, or of their shagreen; while the other two,—its upper and inner sides,—seem to have been developments of the interior osseous walls which covered the endo-skeletal cartilage. Nor is it unworthy of notice, that the reptile fishes of the period had their *ichthyic* teeth ranged along the edge of an exterior *dermal* plate which covered the outer side of the jaw; whereas their *reptile* teeth were planted on a plate, apparently of interior development, which covered its upper edge. It is further worthy of remark, that while the teeth of the dermal plate,—themselves also dermal,—seem as if they had grown out of it, and formed part of it,—just as the teeth of the placoids grow out of the skin on which they rest,—the *reptile* teeth within rested in shallow pits,—the first faint indications of true sockets.

That space included within the arch formed by the sweep of the under jaws, which we find occupied in the osseous fishes by the hyoid bones and the branchiostegous rays, was filled up externally, in the Dipterians and Celacanths, and in at least two genera of Cephalaspians, by dermal plates; in some genera, such as the *Diplopterus*, by three plates; in others, such as the *Holoptychius* and *Glyptolepis*, by two; and in the *Asterolepis*, as we shall afterwards see, by but a single plate. In the *Osteolepis* these plates were increased to five in number, by the little plates 14, 14 (fig. 14), which, however, may have

Fig. 14.

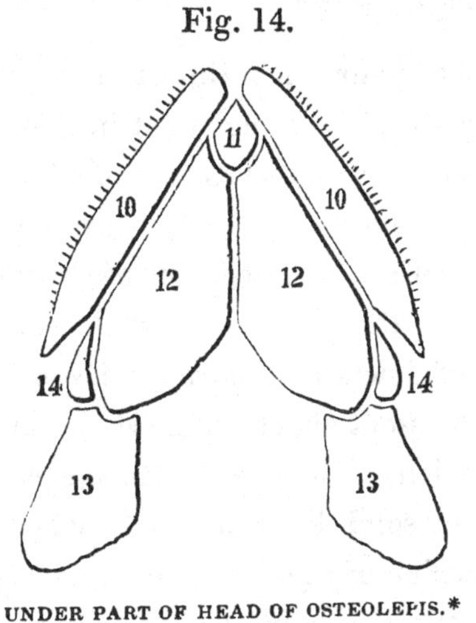

UNDER PART OF HEAD OF OSTEOLEPIS.*

* The jaws (10, 10) which exhibit in the print their greatest breadth, would have presented in the animal, seen from beneath,

been also present in the *Diplopterus*, though my specimens fail to show them. The general arrangement was of much elegance,—an elegance, however, which, in the accompanying restorations, the dislocation of the free plates, drawn apart to indicate their detached character, somewhat tends to obscure. But the position of the eyes must have imparted to the animal a sinister, reptile-like aspect. The profile (fig. 15), the

Fig. 15.

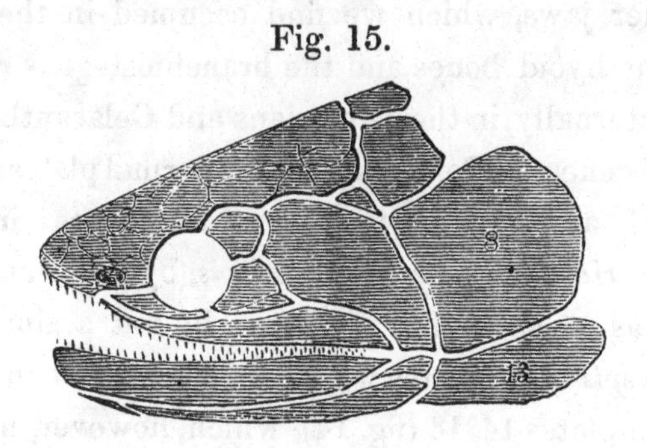

HEAD OF OSTEOLEPIS, SEEN IN PROFILE.

result, not of a chance-drawn outline, arbitrarily filled up, but produced by the careful arrangement in their proper places of actually existing plates, serves to show how perfectly the dermo-skeletal parts of the creature were developed. Some of the animals with which we are best acquainted, if represented by but their cuticular skeleton, would appear simply as sets of hoofs and horns. Even the tortoise or pengolin would present about the head and limbs their gaps and missing portions; but the dermo-skeleton of the *Osteolepis*, composed of solid bone, and burnished with enamel, exhibited the outline of the fish entire, and, with the exception of the eye, the filling up of all its external parts. Pre-

---

their narrow under edges, and have nearly fallen into the line of the sub-opercular plates (13, 13).

senting outside, in its original state, no fragment of skin or membrane, and with even its most flexible organs sheathed in enamelled bone, the *Osteolepis* must have very much resembled a fish carved in ivory; and, though so effectually covered, it would have appeared, from the circumstance, that it wore almost all its bone outside, as naked as the human teeth.

The cranial buckler of the *Diplopterus* (fig. 16) somewhat

Fig. 16

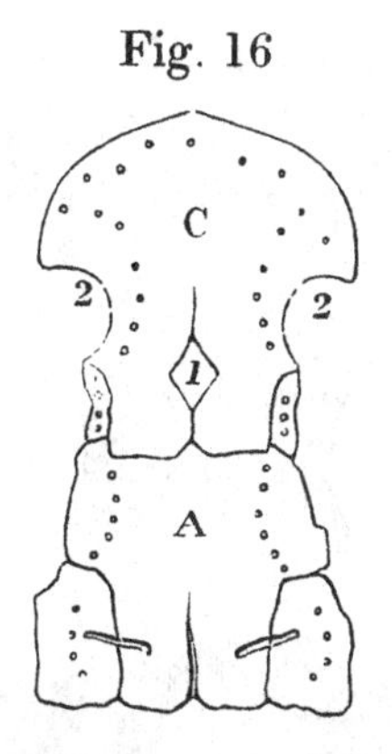

CRANIAL BUCKLER OF DIPLOPTERUS.

resembled that of its fellow-dipterian the *Osteolepis*, but exhibited greater elegance of outline. My first perfect specimen, which I owe to the kindness of Mr John Miller of Thurso, an intelligent geologist of the north, reminded me, as it glittered in jet-black enamel on its ground of pale gray, of those Roman cuirasses which one sees in old prints, impaled on stakes, as the central objects in warlike trophies formed of spoils taken in battle. The rounded snout represented the chest and shoulders, the middle portion the waist, and the expansion at the nape the piece of dress attached, which, like the Highland kilt, fell adown the thighs. The addition of a fragment of a sleeve, suspended a little over the eye orbits, 2, 2, seemed all that was necessary in order to render the resemblance complete. But as I disinterred the buried edges of the specimen with a graver, the form, though it grew still more elegant, became less that of the ancient coat of armour ; the snout expanded into a semicircle ; the eye orbits gradually deepened ; and the entire fossil became not particularly like anything but the thing it once was,— the cranial buckler of the *Diplopterus*. The print (fig. 17)

Fig. 17.

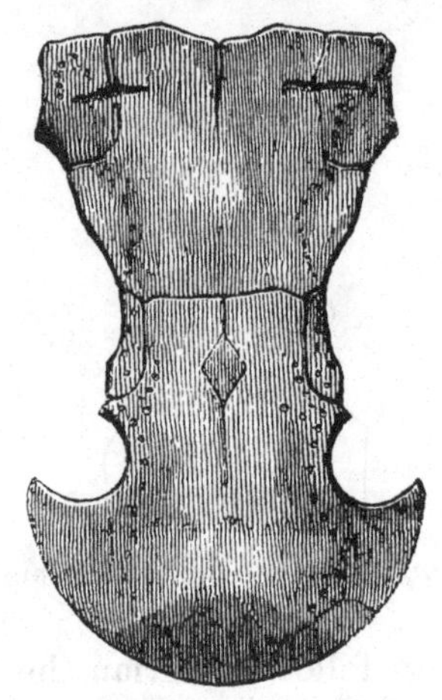

CRANIAL BUCKLER OF DIPLOPTERUS.

exhibits its true form. It consists of two main divisions, occipital (A) and frontal (C, fig. 16) ; and in each of these we find a pair of smaller divisions, with what seem to be indications of yet further division, marked, not by lines, but by dots ; though I have hitherto failed to determine whether the plates which these last indicate possess their independent centres of ossification. Not unfrequently, however, has the comparative anatomist to seek the analogues of two bones in one ; nor is it at least *more* difficult to trace in the faint divisions of the cranial buckler of the *Diplopterus*, the homologues of the occipital, frontal, parietal, mastoid, and nasal bones, than to recognise the representatives of the carpals of the middle and ring finger in man, in the cannon bone of the fore leg of the ox. I may mention in passing, that the little central plate of the frontal division (1, fig. 16), which so nearly corresponds with that of the *Osteolepis*, occurred, though with considerable variations of form and homology, and some slight difference of position, in all the ganoids of the Old Red Sandstone whose craniums were covered with an osseous buckler, and that its place was always either immediately between the eyes or a very little over them. Its never-failing recurrence shows that it must have had *some* meaning, though it may be difficult to say what. In the *Coccosteus* it takes the form of the male dovetail, which united the nasal plate or snout to the plate representative of the superior frontal. Of the cartilaginous box which formed the interior skull of either *Osteolepis* or *Diplopterus*, or, with but one exception, of the interior skulls of any of their contemporaries, no trace, as I have said, has yet been detected. The solitary exception in the case is, however, one of singular interest.

Fig. 18.

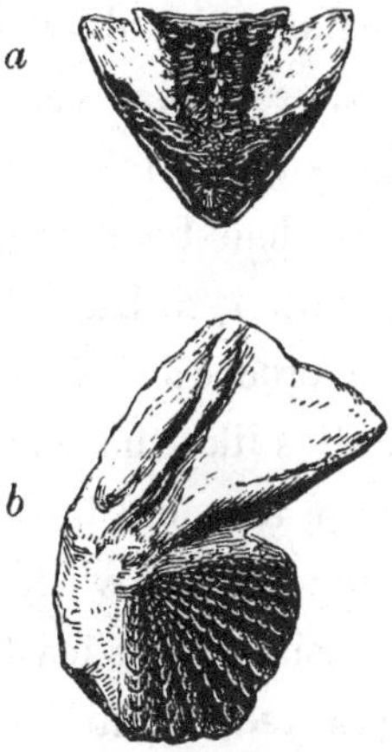

a

b

a, *Palatal dart-head.*
b, *Group of palatal teeth.*

In a collection of miscellaneous fragments sent me by Mr Dick from the rocks of Thurso, I detected patches of palatal teeth ranged in nearly the quadratures of circles, and which radiated outwards from the rectangular angle or centre (fig. 18, *b*). And with the patches there occurred plates exactly resembling the barbed head of a dart (*a*), with which

I had been previously acquainted, though I had failed to determine their character or place. The excellent state of keeping of some of Mr Dick's specimens now enabled me to trace the patches with the dart-head, and several other plates, to a curious piece of palatal mechanism, ranged along the base of a ganoid cranium, covered externally by a brightly enamelled buckler, and to ascertain the order in which patches and plates occurred. And then, though not without some labour, I succeeded in tracing the buckler with which they were associated to the *Dipterus*,—a fish which, though it has engaged the attention of both Cuvier and Agassiz, has not yet been adequately restored. It is on an ill-preserved Orkney specimen of the cranial buckler of this ganoid that the ichthyologist has founded his genus *Polyphractus;* while groupes of its palatal teeth from the Old Red of Russia he refers to a supposed placoid,—the *Ctenodus.* But in the earlier stages of palæontological research, mistakes of this character are wholly unavoidable. The palæontologist who did avoid them would be either very unobservant, or at once very rash and very fortunate in his guesses. If, ere an entire skeleton of the *Ichthyosaurus* had turned up, there had been found in different localities, in the Liasic formation, a beak like that of a porpoise, teeth like that of a crocodile, a head and sternum like that of a lizard, paddles like those of a cetacean, and vertebræ like those of a fish, it would have been greatly more judicious, and more in accordance with the existing analogies, to have erected, provisionally at least, places specifically, or even generically separated, in which to range the separate pieces, than to hold that they had all united in one anomalous genus ; though such was actually the fact. And Agassiz, in erecting three distinct genera out of the fragments of a single genus, has in reality acted at once more prudently and more intelligently than if he had avoided the error by rashly uniting parts which in their separate state indicate no tie of connection.

The cranial buckler of the *Dipterus* (fig 19.), was, like

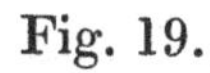

Fig. 19.

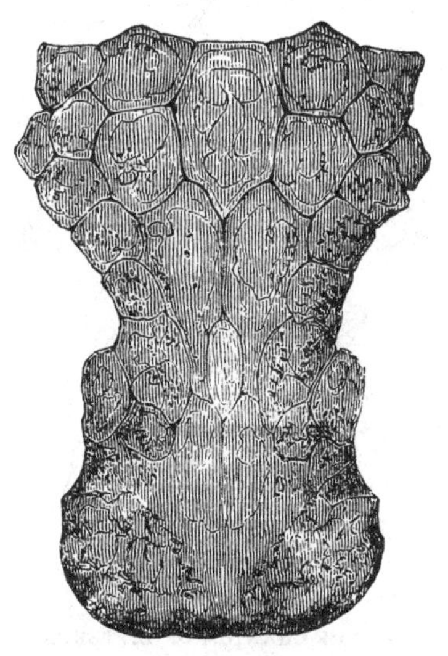

CRANIAL BUCKLER of DIPTERUS.

that of the *Diplopterus*, of great beauty. In some of the finer specimens, we find the enamel ornately tatooed, within the more strongly-marked divisions, by delicately traced lines, waved and bent, as if upon the principle of Hogarth ; and though the lateral plates are numerous and small, and defy the homologies, we may trace in those of the central line, from the snout to the nape, what seem to be the representatives of the frontal, parietal, and occipital bones,—the parietals ranging, as in the skull of the carp and in that of most of the mammals, in their proper place in the medial line. But the under surface of the cranium, armed, as on the upper surface, with plates of bone, exhibited an arrangement still more peculiar (fig. 20). Its rectangular patches of palatal teeth, its curious dart-like bone, placed imme-

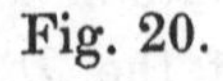

Fig. 20.

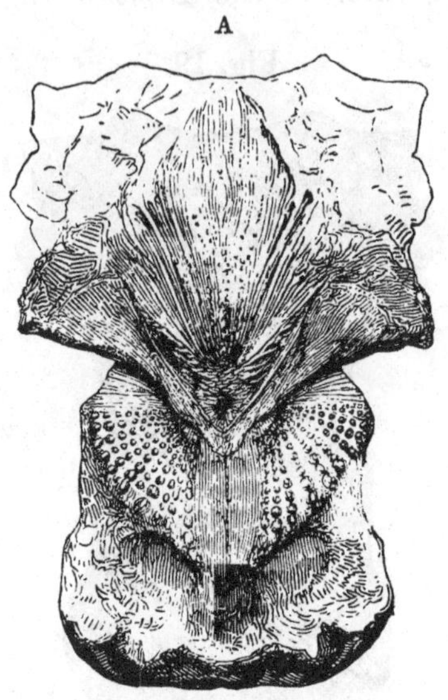

BASE OF CRANIUM OF DIPTERUS.

diately behind these, and attached, as the dart-head is attached to the handle, to a broad lozenge-shaped plate, with two strong osseous processes projecting on either side, forms such a *tout ensemble* as is unique among fishes. Even here, however, there may be traced at least a shade of homological resemblance to the bones which form the base of the osseous skull. The single lozenge-shaped plate (A), with its dart-head, occupies the place of the basi-occipital bone ; the posterior portion of the vomer seems represented by a strong bony ridge, extending towards the snout ; two separate bones, each bearing one of the angular patches of teeth, corresponds to the sphenoid bone and its alæ; and attached laterally to each of these there is the strong projecting bone, on which the lower jaw appears to have hinged, and which apparently represents the lower part of the temporal bone. Not less singular was the form of the creature's under jaw (fig. 21).

Fig. 21.

UNDER JAW OF DIPTERUS.

I know no other fish-jaw, whether of the recent or the extinct races, that might be so readily mistaken for that of a quadruped. It exhibits not only the condyloid, but also the coronoid processes ; and, save that it broadens on its upper edges, where in mammals the grinders are placed, so as to furnish field enough for angular patches of teeth, which correspond with the angular patches in the palate, it might be regarded, found detached, as at least a reptilian, if not mammalian, bone. The disposition of the palatal teeth of the *Dipterus* will scarce fail to remind the mechanist of the style of grooving resorted to in the formation of mill-stones for the grinding of flour ; nor is it wholly improbable that, in correspondence with the rotatory motion of the stones to which the grooving is specially adapted, jaws so hinged may have possessed some such power of lateral motion as that exemplified by the human subject in the use of the molar teeth.

The protection afforded by the osseous covering of both the upper and under surface of the cranium of this ichthyolite has resulted, in several instances, in the preservation, though always in a greatly compressed state, of the cranium itself, and the consequent exhibition of two very important cranial cavities, the brain-pan proper, and the passage through which the spinal cord passed into the brain. In the sturgeon the brain occupies nearly the middle of the head ; and there is a considerable part of the occipital region traversed by the spine in a curved channel, which, seen in profile, appears wide at the nape, but considerably narrower where it enters the brain-pan, and altogether very much resembling the interior of a miniature hunting-horn. And such exactly was the arrangement of the greater cavities in the head of the *Dipterus.* The portion of the cranium which was overlaid by what may be regarded as the occipital plate was traversed by a cavity shaped like a Lilliputian bugle-horn ; while the hollow in which the brain was lodged lay under the two parietal plates, and the little elliptical plate in the centre. The accompanying print (fig. 22),

Fig. 22.

LONGITUDINAL SECTION OF HEAD OF DIPTERUS.

though of but slight show, may be regarded by the reader with some little interest, as a not inadequate representation of the most ancient brain-pan on which human eye has yet looked,— as, in short, the type of cell in which, myriads of ages ago, in at least one genus, that mysterious substance was lodged, on whose place and development so very much in the scheme of creation was destined to depend. The specimen from which the figure is taken was laid open laterally by chance exposure to the waves on the shores of Thurso ; another specimen, cut longitudinally by the saw of the lapidary, yields a similar section, but greatly more compressed in the cavities; on which, of course, as unsupported hollows, the compression to which the entire cranium had been exposed chiefly acted. When the top and bottom of a box are violently forced together, it is the empty space which the box encloses that is annihilated in consequence of the violence.

It is deserving of notice, that the analogies of the cranial cavities in this ancient ganoid should point so directly on the cranial cavities of that special ganoid of the present time which unites a true skull of cartilage to a dermal skull of osseous plates,—a circumstance strongly corroborative of the general evidence, negative and positive, on which I have concluded that the true skulls of the first ganoids were also cartilaginous. It is further worthy of observation, that in all the sections of the cranium of *Dipterus* which I have yet examined, the internal line is continuous, as in the placoids, from nape to snout, and that the true skull presents no trace of those cerebral vertebræ of which skulls are regarded by Oken and his disciples as developments. Historically at least, the progress of the ichthyic head seems to have been a progress from simple cartilaginous boxes to cartilaginous boxes covered with osseous plates, that performed the functions, whether active or passive, of internal bones; and then from external plates to the interior bones which the plates had previously represented, and whose proper work they had done.

The principle which rendered it necessary that the divisions which exist in the dermal skulls of the first ganoids should so closely correspond with the divisions which exist in the internal skulls of the osseous fishes of a greatly later period, does not seem to lie far from the surface. Of the solid parts of the ichthyic head, a certain set of pieces afford protection to the brain and cerebral nerves, and to some of the organs of the senses, such as those of seeing and hearing; while another certain set of pieces constitute the framework through which an important class of functions, manducatory and respiratory, are performed. The protective bones of merely passive function are fixed, whereas the bones of active function, such as the jaws, the osseous framework of the opercules, and the hyoid bones, are to the necessary extent free, *i. e.* capable of independent motion. Of course, the detached character necessary to the free cerebral bones would be equally necessary in cerebral plates united dermally to the pieces of the cartilaginous framework, which performed in the ancient fish the functions of these free bones. And hence jaw plates, opercular plates, and hyoid plates, whose homological relation with recent jaws and opercular and hyoid bones cannot be mistaken. They were operative in performing identical mechanical functions, and had to exist, in consequence, in identical mechanical conditions. And an equally simple, though somewhat different principle, seems to have regulated the divisions of the fixed cranial bucklers of the Old Red ganoids, and to have determined their homologies with the fixed cerebral bones of the osseous fishes.

These cranial bucklers, extending from nape to snout, protected the exposed upper surface of the cartilaginous skull, and conformed to it in shape, as a helmet conforms to the shape of the head, or a breast-plate to the shape of the chest. And as the cartilaginous heads resembled in general outline the osseous ones, the buckler which covered their upper surface resembled in general outline the upper surface of the osseous skull. It was in no case entirely a flat plate; but in every species rounded over the snout, and in most species at the sides ; and so, in order that its characteristic proportions might be preserved throughout the various stages of growth in the head which it covered, it had to be formed from several distinct centres of ossification, and to extend in area around the edges of the plates originated from these. The workman finds no difficulty in adding to the size of a piece of straight wall, whether by heightening or lengthening it ; but he cannot add to the size of a dome or arch, without first taking it down, and then erecting it anew on a larger scale. In the domes and arches of the animal kingdom, the problem is solved by building them up of distinct pieces, few or many according to the demands of the figure which they compose, and rendering these pieces capable of increase along their edges. It is on this principle that the Cystidea, the Echinidæ, the Chelonian carpace and plastron, and the skulls of the osseous vertebrata, are constructed. It is also the principle on which the cranial bucklers of the ancient ganoids were formed.* And from the general resemblance in figure of these bucklers to the upper surface of the osseous skull, the separate parts necessary for the building up of the one were anticipated, by many ages, in the building up of the other ; just as we find external arches of stone

* In all probability it is likewise the principle of the placoid skull. The numerous osseous points by which the latter is encrusted, each capable of increase at the edges, seem the minute bricks of an ample dome. It is possible, however, that new points may be formed in the interstices between the first formed ones, as what anatomists term the *triquetra* or *Wormiana* form between the serrated edges of the lambdoidal suture in the human skull ; and that the osseous surface of the cerebral dome may thus extend, as the dome itself increases in size, not through the growth of the previously existing pieces,—the minute bricks of my illustration,—but through the addition of new ones. Equally, in either case, however, that essential difference between the pla-

which were erected two thousand years ago, constructed on the same principle, and relatively of the same parts, as internal arches of brick built in the present age. Doubtless, however, with this mechanical necessity for correspondence of parts in the formation of corresponding erections, there may have mingled that regard for typical resemblance which seems so marked a characteristic of the *style*, if I may so express myself, in which the Divine Architect gives expression to his ideas. The external osseous buckler He divided after the general pattern which was to be exemplified, in latter times, in the divisions of the internal osseous skull; as if in illustration of that "ideal exemplar" which dwelt in his mind from eternity, and on the palpable existence of which sober science has based deductions identical in their scope and bearing with some of the sublimest doctrines of the theologian. "The recognition," says Professor Owen, "of an ideal exemplar for the vertebrated animals, proves that the knowledge of such a being as man existed before man appeared; for the Divine mind which planned the archetype also foreknew all its modifications. The archetypal idea was

---

coid skull and the placoid vertebra, to which I have referred, appears to hinge on the circumstance, that while the osseous nucleus of each vertebral centrum could form, in even its most complicated shape, from a *single* point, the osseous walls of the cranium had to be formed from *hundreds*. The accompanying diagram serves to show after what manner the vertebral centrum in the Ray enlarges with the growth of the animal, by addition of bony matter external to the point in the middle, at which ossification first begins. The horizontal lines indicate the lines of increment in the two internal cones which each centrum comprises, and the vertical ones the lines of increment in the lateral pillars.

Fig. 23.

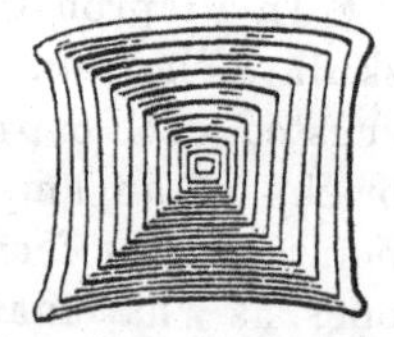

SECTION OF VERTEBRAL CENTRUM OF THORNBACK.

manifested in the flesh, under divers such modifications, upon this planet, long prior to the existence of those animal species that actually exemplify it."

But while we find place in that geological history in which every character is an organism, for the "ideal exemplar" of Professor Owen, we find *no* place in it for the vertebræ-developed skull of Professor Oken. The true genealogy of the head runs in an entirely different line. The nerves of the cerebral senses did not, we find, originate cerebral-vertebræ, seeing that the heads of the first and second geologic periods had their cerebral nerves, but *not* their cerebral vertebræ ; and that what are regarded as cerebral-vertebræ appear for the first time, not in the early fishes, but in the reptiles of the Coal formation. That line of succession through the fish, indicated by the Continental assertor of the development hypothesis, is a line cut off. All the existing evidence conspires to show that the placoid heads of the Silurian system were, like the placoid heads of the recent period, mere cartilaginous boxes; and that in the succeeding system there existed ganoidal heads, that to the internal cartilaginous box added external plates of bone, the homologues, apparently,—so far at least as the merely cuticular could be representative of the endo-skeletal,—of the opercular, maxillary, frontal, and occipital bones in the osseous fishes of a long posterior period,—fishes that were not ushered upon the scene until after the appearance of the reptile in its highest forms, and of even the marsupial quadruped.

THE ASTEROLEPIS, ITS STRUCTURE, BULK, AND ASPECT.

WITH the reader, if he has accompanied me thus far, I shall now pass on to the consideration of the remains of the *Asterolepis*. Our preliminary acquaintance with the cerebral peculiarities of a few of its less gigantic contemporaries will be found of use in enabling us to determine regarding a class of somewhat resembling peculiarities which characterized this hugest ganoid of the Old Red Sandstone.

The head of the *Asterolepis*, like the heads of all the other Celacanths, and of all the Dipterians, was covered with osseous plates,—its body with osseous scales; and, as I have already had occasion to mention, it is from the star-like tubercles by which the cerebral plates were fretted that M. Eichwald bestowed on the creature its generic name. Agassiz has even erected species on certain varieties in the pattern of the stars, as exhibited on detached fragments; but I am far from being satisfied that we are to seek in their peculiarities of style the characters by which the several species were distinguished. The stellar form of the tubercle seems to have been its normal or most perfect form, as it was also, with certain modifications, that of the tubercle of the *Coccosteus* and *Pterichthys*; but its development as a complete star was comparatively rare: in most cases the tubercles existed without the rays,—frequently in the insulated pap-like shape, but not rarely confluent, or of an elongated or bent form; and when to these the characteristic rays were added, the stars produced were of a rather eccentric order,—stars somewhat resembling the shadows of stars seen in water. Individual specimens have already been found, on which, if we recognise the form of the tubercle as a specific character, several species might be erected. The accompanying wood-cut (fig. 24) represents, from a Thurso specimen, what seems to be the true normal pattern of these cerebral carvings. Seen in profile (*b*), the tubercles resemble little hillocks, perforated at their bases by single lines of thickly-set caves; while seen from above (*a*), the narrow piers of bone by which the caves are divided take the form of rays. The reader will scarce fail to recognise in this print the coral *Monticularia* of Lamarck, or to detect, in at least the profile, the peculiarity which suggested the name.

Fig. 24.

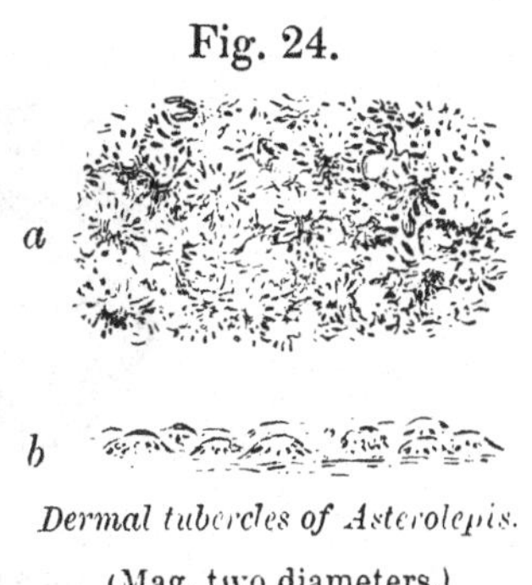

Dermal tubercles of Asterolepis.

(Mag. two diameters.)

The scales which covered the creature's body (fig. 25) were, in proportion to its size, considerably smaller and thinner than those of the *Holoptychius*, which, however, they greatly resemble in their general style of sculpture. Each, on the lower part of its exposed field, was, we see, fretted by longitudinal anastomosing ridges, which, in the upper part, break into detached angular tubercles, placed with the apex downwards, and hollowed, leaf-like, in the centre; while that covered portion which was overlaid by the scales immediately above we find thickly pitted by miscroscopic hollows, that give to this part of the field, viewed under a tolerably high

Fig. 25.

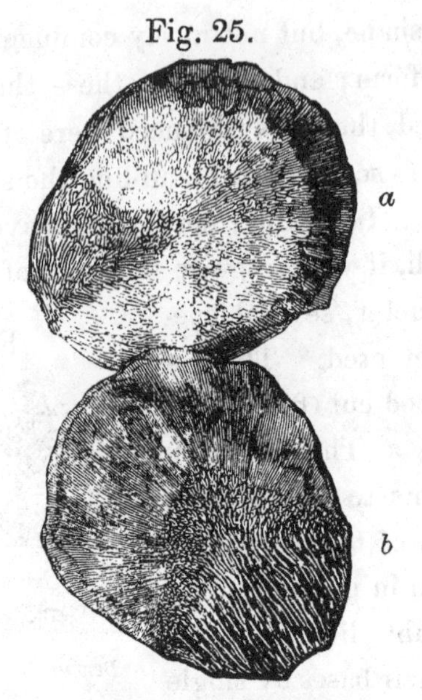

SCALES OF ASTEROLEPIS.
(Nat. size)

a. *Inner surface of scale.* b *Exterior surface.*

Fig. 26.

PORTION OF CARVED SURFACE OF SCALE.
(Mag. four diameters.)

magnifying power, a honeycombed appearance. The central and lower parts of the interior surface of the scale (*a*) are in most of the specimens irregularly roughened; while a broad, smooth band, which runs along the top and sides, and seems to have furnished the line of attachment to the creature's body, is comparatively smooth. The exterior carvings, though they demand the assistance of the lens to see them aright, are of singular elegance and beauty; as perhaps the accompanying wood-cut (fig. 26), which gives a magnified view of a portion of the scale immediately above (*b*), from the middle of the honeycombed field on the right side, to where the anastomosing ridges bend gracefully in their descent, may in some degree serve to show. I have seen a richly inlaid coat of mail, which was once worn by the puissant Charles the Fifth ; but its elaborate carvings, though they belonged to the age of Benvenuto Cellini, were rude and unfinished, compared with those which fretted the armour of the *Asterolepis*.

The creature's cranial buckler, which was of great size and strength, might well be mistaken for the carpace of some Chelonian fish of no inconsiderable bulk. The cranial bucklers of the larger Dipterians were ample enough to have covered the corresponding part in the skulls of our middle-sized market-fish, such as the haddock and whiting ; the buckler of a *Coccosteus* of the extreme size would have covered, if a little altered in shape, the upper surface of the skull of a cod ; but the cranial buckler of *Asterolepis*, from which the accompanying woodcut was taken (fig. 27), would have considerably more than covered the corresponding part in the skull of a large horse ; and I have at least one specimen in my collection which would have fully covered the front skull of an elephant. In the smaller specimens, the buckler somewhat resembles a labourer's shovel divested of its handle, and sorely rust-eaten along its lower or cutting edge. It consisted of plates, connected at the edges by flat squamous sutures, or, as a joiner might perhaps say, *glued* together in *bevelled* joints. And in consequence of this arrangement, the same plates which seem broad on the exterior surface appear comparatively narrow on the interior one, and *vice versa:* the occipital plate (*a*), which, running from the nape along the centre of the buckler, occupies so considerable a space on its outer surface, exhibit inside a superficies reduced at least one-half. Like nine-tenths of its contemporaries, the *Asterolepis* exhibits the little central plate be-

Fig. 27.

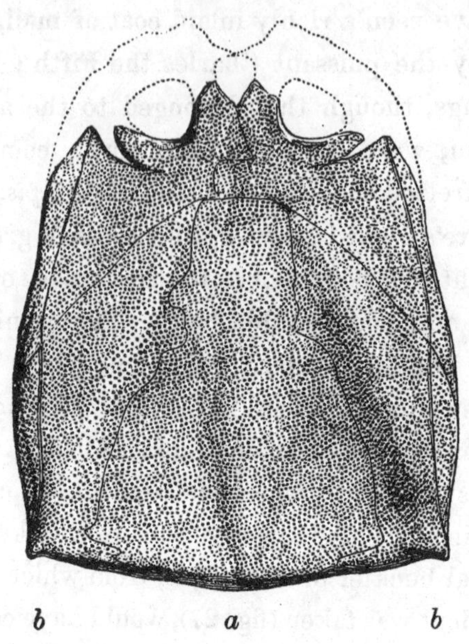

CRANIAL BUCKLER OF ASTEROLEPIS.

(One-fifth nat. size, linear.)

-tween the eyes; but the eye-orbits, unlike those of the *Coccosteus*, and of all the Dipterian genera, which were half-scooped out of the cranial buckler, half-encircled by detached plates, were placed completely within the field of the buckler,—a circumstance in which they resemble the eye-orbits of the *Pterichthys*, and, among existing fish, those of the sea-wolf. The characteristic is also a distinctive one in Cuvier's second family of the Acanthopterygii,—the "fishes with hard cheeks." A deep line immediately over the eyes, which, however, indicated no suture, but seems to have been merely ornamental, forms a sort of rudely tatooed eyebrow; the marginal lines parallel to the lateral edges of the buckler were also mere tatooings; but all the others indicated joints which, though more or less anchylosed, had a real existence. So flat was the surface, that the edge of a ruler rests upon it, in my several specimens, both lengthwise and across; but it was traversed by two flat ridges, which, stretching from the corners of the latero-posterior, *i. e.* parietal, plates (*b*, *b*), converged at the little plate between the eyes; while along the centre of the depressed angle which they formed, a third ridge, equally flat with the others, ran towards the same point of convergence from the nape. The three ridges, when strongly relieved by a slant light, resemble not inadequately an impression, on a large scale, of the Queen's broad arrow.

Fig. 28.

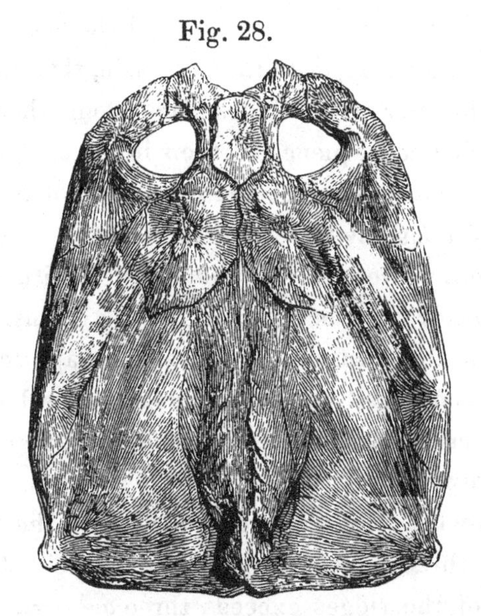

INNER SURFACE OF CRANIAL BUCKLER OF ASTEROLEPIS.
(One-fifth nat. size, linear.)

The inner surface of the cranial buckler of *Asterolepis* (fig. 28),—that which rested on the cartilaginous box which formed the creature's interior skull,—stands out in bolder relief from the stone than its outer surface, and forms a more picturesque object. Like the inner surfaces of the bucklers of *Coccosteus* and *Pterichthys,* but much more thickly than these, it was traversed by minute channelled markings, somewhat resembling those striæ which may be detected in the flatter bones of the ordinary fishes, and which seem in these to be mere interstices between the osseous fibres. And in the plates, as in the bones, they radiate from the centres of ossification, which are comparatively dense and massy, towards the thinner overlapping edges. These radiating lines are equally well marked in the cerebral bones of the human fœtus. The three converging ridges on the outer surface we find on the inner surface also,—the lateral ones a little bent in the middle, but so directly opposite those outside, that the thickening of the buckler which takes place along their line is at least as much a consequence of their inner as of their outer elevation over the general platform. A fourth bar ran transversely along the nape, and formed the cross beam on which the others rested ; for the three longitudinal ridges may be properly regarded as three strong beams, which, extending from the transverse beam at the nape to the front, where they converged like the spokes of a wheel at the nave, gave to the cranial roof a degree of support of which, from its great flatness, it may have stood in need. In cranial bucklers in which the average thickness of the plates does not exceed three *eighth* parts of an inch, their thickness in the centre of the ridges exceeds three *quarters.* The head of the largest crocodile of the existing period is defended by an armature greatly less strong than that worn by the *Asterolepis* of the Lower Old Red Sandstone. Why this ancient ganoid should have been so ponderously helmed we can but doubtfully guess ; we only know, that when nature arms her soldiery, there are assailants to be resisted, and a state of war to be maintained. The posterior central plate, the homologue apparently of the occipital bone, was curiously carved into an ornate massive leaf, like one of the larger leaves of a Corinthian capital, and terminated beneath, where the stem should have been, in a strong osseous knob, fashioned like a pike-head. Two plates immediately over it, the homologues of the superior frontal bone, with the little nasal plate which, perched atop in the middle, lay between the creature's eyes, resembled the head and breasts in the female figure, at least not less closely than those of the "lady in the lobster;" the posterior frontal plates in which the outer and nether half of the eye-orbits were hollowed formed a pair of sweeping wings; and thus in the centre of the buckler we are presented with the figure of an angel, robed and winged, and of which the large sculptured leaf forms the body, traced in a style in no degree more rude than we might expect to see exemplified on the lichen-encrusted shield of some ancient tombstone of that House of Avenel which bore as its arms the effigies of the Spectre Lady. Children have a peculiar knack in detecting such resemblances; and the discovery of the angel in the cranium of the *Asterolepis* I owe to one of mine.

It is on this inner side of the cranial buckler, where there are no such pseudo-joinings indicated as on the external surface, that the homologies of the plates of which it is composed can be best traced. It might be well, however, ere setting one's self to the work of comparison, to examine the skulls of a few of the osseous fishes of our coasts, and to mark how very considerably they differ from one another in their lines of suture and their general form. The cerebral divisions of the conger-eel, for instance, are very unlike those of the haddock or whiting; and the sutures in the head of the gurnard are dissimilarly arranged from those in the head of the perch. And after tracing the general type in the more anomalous forms, and finding, with Cuvier, that in even these the "skull consists of the same bones, though much subdivided, as the skulls of the other vertebrata," we will be the better qualified for grappling with the not greater anomalies which occur in the cranial buckler of the *Asterolepis.* The occipital plate, *A, a, a* (fig. 29), occupies its ordinary place op-

Fig. 29.

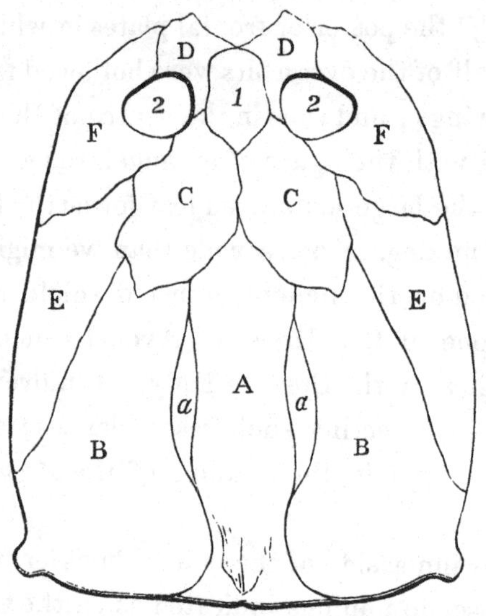

PLATES OF CRANIAL BUCKLER OF ASTEROLEPIS.

posite the centre of the nape ; the two parietals, B, B, rest beside it in their usual ichthyic position of displacement ; the superior frontal we find existing, as in the young of many animals, in two pieces, C, C ; the nasal plate I, placed immediately in advance of it, is flanked, as in the cod, by the anterior frontals, D, D ; the posterior frontals, F, F, which, when viewed, as in the print, from beneath, seem of considerable size, and describe laterally and posteriorly about one-half the eye-orbits, have their area on the exterior surface greatly reduced by the overriding squamose sutures of the plates to which they join ; and lastly, two of these overlying plates, E, E,—which, occurring in the line of the lateral bar or beam, are of great strength and thickness, and lie for two-thirds of their length along the parietals, and for the remaining third along the superior frontals,—represent the mastoid bones. Such, so far as I have been yet able to read the cranial buckler of the *Asterolepis*, seem to be the homologies of its component plates.

There were no parts of the animal more remarkable than its jaws. The under jaws,—for the nether maxillary consisted, in this fish, as in the placoid fishes, and in the quadrupeds generally, of two pieces joined in the middle,—were, like those of the *Holoptychius*, boxes of bone, which enclosed central masses of cartilage. The outer and under sides were thickly covered with the characteristic star-like tubercles ; and along the upper margin or lip there ran a thickly-set row of small broadly-based teeth, planted as directly on the edge of the exterior plate as iron spikes on the upper edge of a gate (fig. 30). Mr Parkinson expresses some wonder,

Fig. 30.

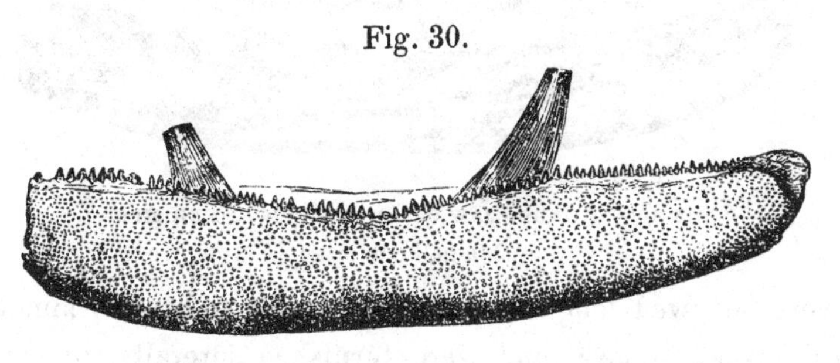

PORTION OF UNDER JAW OF ASTEROLEPIS (OUTER SIDE).

(One-half nat. size.)

in his work on fossils, that, in a fine ichthyolite in the British Museum, not only the *teeth* should have been preserved, but also the *lips*; but we now know enough of the construction of the ancient ganoids to cease wondering. The *lips* were formed of as solid bone as the teeth themselves, and had as fair a chance of being preserved entire; just as the metallic rim of a *cogged* wheel has as fair a chance of being preserved as the metallic *cogs* that project from it. Immediately behind the front row,—in which the teeth present the ordinary ichthyic appearance,—there ran a thinly-set row of huge *reptile* teeth, based on an interior platform of bone, which formed the top of the cartilage-enclosing box composing the jaw. These were at once bent outwards and twisted laterally, somewhat like nails that have been drawn out of wood by the claw of a carpenter's hammer, and bent awry with the wrench (fig. 31). They

Fig. 31.

PORTION OF UNDER JAW OF ASTEROLEPIS (INNER SIDE).
(One-half nat. size.)

were furrowed longitudinally from point to base by minute thickly-set striæ; and were furnished laterally, in most of the specimens, though not in all, with two sharp cutting edges. The reptile had as yet no existence in creation;

but we see its future coming symbolized in the dentition of this ancient ganoid : it, as it were, shows us the *crocodile* lying entrenched behind the fish. The interior structure of these reptile teeth is very remarkable. In the longitudinal section we find numerous cancelli, ranged lengthwise along the outer edges, but much crossed, net-like, within,—greatly more open towards the base than at the point,—and giving place in the centre to a hollow space, occasionally traversed by a few slim osseous partitions. In the transverse section these cancelli are found to radiate from the open centre towards the circumference, like the spokes of a wheel from the nave ; and each spoke seems as if, like Aaron's rod, it had become instinct with vegetative life, and had sprouted into branch and blossom. Seen in a microscope of limited field, that takes in, as in the accompanying print (fig. 32) not more

Fig. 32.

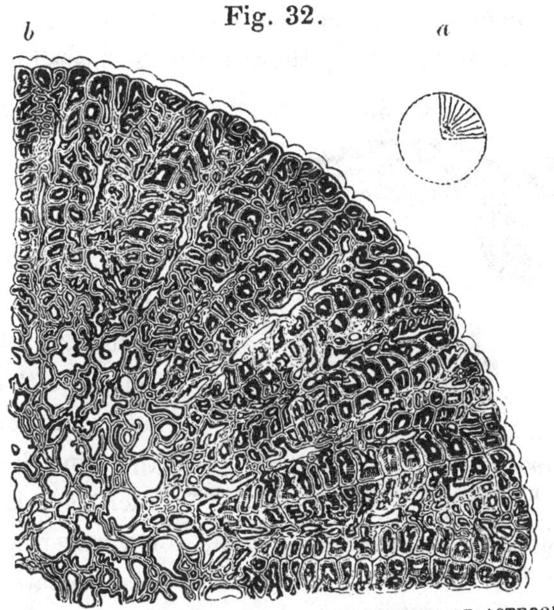

PORTION OF TRANSVERSE SECTION OF REPTILE TOOTH OF ASTEROLEPIS.
a. Nat. size. b *Mag. twelve diameters.*

than a fourth part of the section, the appearance presented is that of a well-trained wall tree. And hence the generic name *Dendrodus*, given by Professor Owen to teeth found detached in the deposits of Moray, when the creatures to which they had belonged were still unknown,—a name, however, which will, I suspect, be found synonymous rather with that of a family than of a genus; for so far as I have yet examined, I find that the dendrodic or tree-like tooth was, in at least the Old Red Sandstone, a characteristic of all the Celacanth family. I may mention, however, as a curious subject of enquiry, that the Celacanths of the Coal Measures seem to have had their reptile teeth formed of pure ivory, —a substance which I have not yet detected among the reptile-fish of the Old Red. Towards the base of the reptile teeth of *Asterolepis*, the interstices between the branches greatly widen, as in the branches of a tree in winter divested of its foliage (fig. 33, *c*); the texture also opens towards the base in the *fish*-teeth outside, in which, however, the pattern in the transverse section is greatly less complex and ornate than that which the reptile teeth exhibits. When cut across near the point, they appear each as a thick ring (*b*), traversed by lines that radiate towards the centre; when cut across about half-way down, they somewhat resemble, seen under a high magnifying power,

Fig. 33.

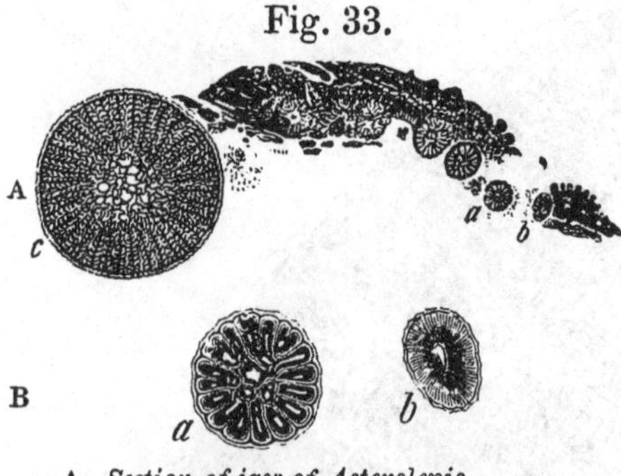

A. *Section of jaw of Asterolepis.*
c. *Reptile tooth as shown in section.*
a, b, & c. *Row of ichthyic teeth in dermal plate of jaw.*
B. *Magnified representatives of ichthyic teeth, a and b, in* A.

those cast-iron wheels on which the engineer mounts his railway carriages (*a*). In the longitudinal section their line of junction with the jaw is marked by numerous openings, but by no line of division, and they appear as thickly dotted by what were once canaliculi, or life points, as any portion of the dermal bone on which they rest.

It seems truly wonderful, when one considers it, to what minute and obscure ramifications that variety of pattern which nature so loves to maintain is found to descend. It descends in the fishes, both recent and extinct, to even the microscopic structure of their teeth ; and we find, in consequence, not less variety of figure in the sliced fragments of the teeth of the ichthyolites of a single formation, than in the carved blocks of an extensive calico print-yard. Each *species* has its own distinct pattern, as if, in all the individuals of which it consisted, the same block had been employed to stamp it ; and each *genus* its own general *type* of pattern, as if the same radical idea, variously altered and modified, had been wrought upon in all. In the *Dendrodic* (Celacanth ?) family, for instance, it is the radical type, that from a central nave there should radiate, spoke-like, a number of arborescent branches ; but in the several genera and species of the family, the branches belong, if I may so express myself, to different shrubs, and present dissimilar outlines. It has appeared to me, that at least a *presumption* against the transmutation of species might be based on those inherent peculiarities of structure which are thus found to pervade the entire texture of the framework of animals. If we find erections differing from one another merely in external form, we have no difficulty in conceiving how, by additions and alterations, they might be brought to exhibit a perfect uniformity of plan and aspect : *transmutation,—development,—progression,*—(if one may use such terms),—seem possible in such circumstances. But if the buildings differ from each other, not only in external form, but also in every brick and beam, bolt and nail, no mere scheme of external alteration could ever induce a real resemblance. Every brick would have to be taken down, and every beam and bolt removed. The problem could not be wrought by the remodelling of an old house : the only mode of solving it would be by the erection of a new one.

Of the upper maxillary bones of the *Asterolepis*, I only know that a considerable fragment of one of the pieces, recognised as such by Agassiz, has been found in the neighbourhood of Thurso by Mr Dick, unaccompanied, however, by any evidence respecting its place or function. It exhibits none of the characteristic tubercles of the dermal bones, and no appearance of teeth ; but is simply a long bent bone, resembling somewhat less than the half of an ancient bow of steel or horn,—such a bow as that which Ulysses bended in the presence of the suitors. By some of the Russian geologists this bone was at first regarded as a portion of the arm or wing of some gigantic *Pterichthys*. In the accompanying print (fig. 34), I have borrowed the general outline from that

Fig. 34.

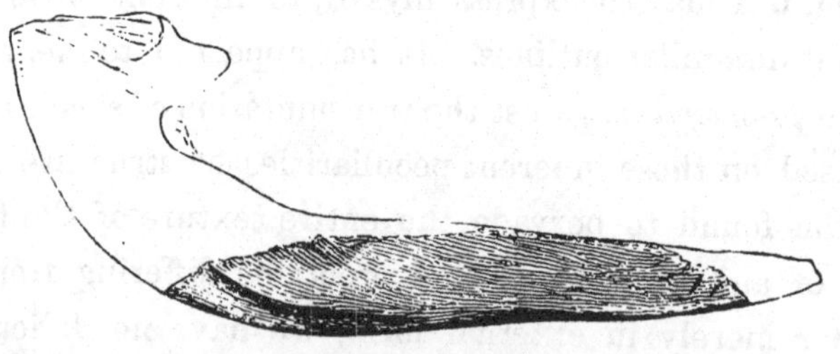

MAXILLARY BONE ?
(One-fourth nat. size, linear.)

of a specimen of Professor Asmus, of which a cast may be seen in the British Museum ; while the shaded portion represents the fragment found by Mr Dick. The intermaxillary bones, like the dermal plates of the lower jaw, were studded by star-like tubercles, and bristled thickly along their lower edges with the ichthyic teeth, flanked by teeth of the reptilian character. The opercules of the animal consisted, as in the sturgeon, of single plates (fig. 35) of great massiveness and size, thickly tubercled outside, without trace of joint or suture, and marked on their under surface by channelled lines, that radiate, as in the other plates, from the centre of ossification. That space along the nape which intervened between the opercules, was occupied, as in the *Dipterus* and *Diplopterus*, by three plates, which covered rather the anterior portion of the body than the posterior portion of the head, and which, in the restoration of *Osteolepis* (fig. 13), appear as the plates, 9, 9, 9. I can say scarce anything regarding the lateral plates which lay between the intermaxillaries and the cranial buckler, and which exist in the *Osteolepis*, fig. 13, as the plates 2, 4, 5, 6, and 7 ; nor do I know how the snout terminated, save that in a very imperfect specimen it exhibits, as in the *Diplopterus* and *Osteolepis*, a rounded outline, and was set with teeth.

Fig. 35.

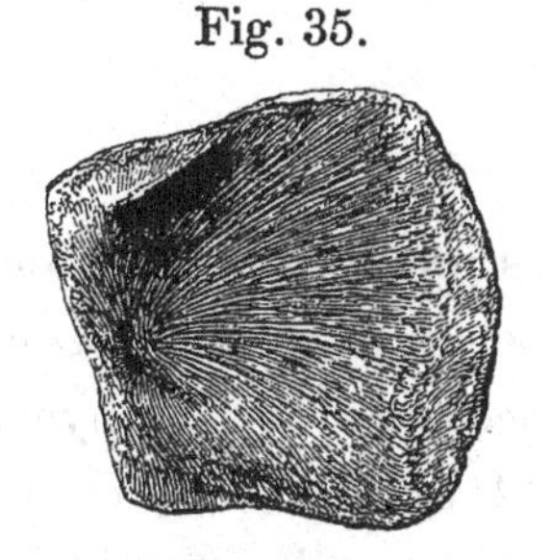

INNEE SURFACE OF OPERCULUM OF ASTEROLEPIS.
(One-fifth nat. size, linear.)

That space comprised within the arch of the lower jaws, in which the hyoid bone and branchiostegous rays of the osseous fishes occur, was filled by a single plate of great size and strength, and of singular form (fig. 36); and to this plate, existing as a steep ridge running along the centre of the interior surface, and thickening into a massy knob at the anterior termination, that nail-shaped organism, which I have described

Fig. 36.

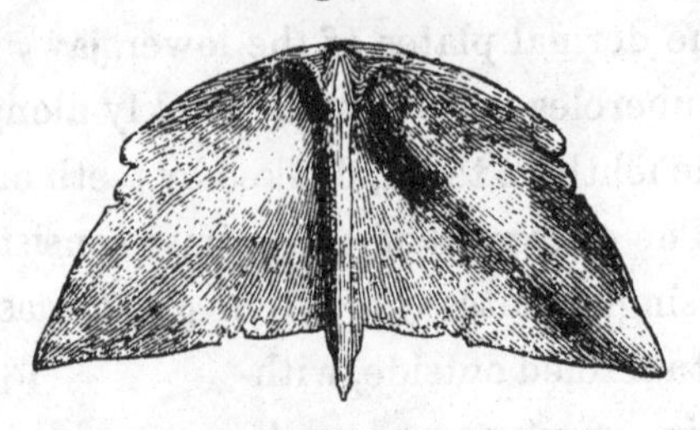

HYOID PLATE.

(One-ninth nat. size, linear.)

as one of the most characteristic bones of the *Asterolepis*, belonged. In the *Osteolepis*, the space corresponding to that occupied by this hyoid plate was filled, as shown in fig. 14, by five plates of not inelegant form ; and the divisions of the arch resembled those of a small Gothic window, in which the single central mullion parts into two branches atop. In the *Holoptychius* and *Glyptolepis* there were but two plates ; for the central mullion, *i. e.* line of division, did not branch atop ; and in the *Asterolepis*, where there was no line of division, the strong nail-like bone occupied the place of the central mullion. The hyoidal armature of the latter fish was strongest in the line in which the others were weakest. Each of the five hyoid plates of the *Osteolepis*, or of the two plates of the *Glyptolepis* or *Holoptychius*, had its own centre of ossification ; and in the single plate of *Asterolepis*, the centre of ossification, as shown by the radiations of the fibre, was the *nail*-head. This head, placed in immediate contact with the strong boxes of bone which composed the under jaw, just where their central joining occurred, seems to have lent them a considerable degree of support, which at such a juncture may have been not unnecessary. In some of the nail-heads, belonging, it is probable, to a different species of *Asterolepis* from that in which the nail figured in page 7, and the plate in the opposite page, occurred,—for its general form is different (fig. 37),—there appear well-marked ligamentary impressions, closely resembling that little spongy pit in the head of the human thigh-bone to which what is termed the round ligament is attached. The entire hyoid-plate, viewed on its outer side, resembles in form the hyoid-bone,—or cartilage rather,—of the spotted dog-fish (*Scyllium stellare*); but its area was at least a hundred times more extensive than in the largest *Scyllium*, and, like all the dermal plates of the *Asterolepis*, it was thickly fretted by the characteristic tubercles. In the Ray, as in the Sharks, the piece of thin cartilage of which this plate seems the homologue, is a flat, semi-transparent disk; and there is no part of the animal in which the progress of those bony molecules which encrust the internal framework may be more distinctly traced, as if in the act of creeping over what they cover, in slim threads or shooting points,—and much resembling new ice creeping in a frosty evening over the surface of a pool.

Fig. 37.

NAIL-LIKE BONE OF HYOID PLATE.
(One-half nat. size.)

That suite of shoulder-bones that in the osseous fishes forms the belt or frame on which the opercules rest, and furnishes the base of the pectorals, was represented in the *Asterolepis*, as in the sturgeon, by a ring of strong osseous plates, which, in one of the two species of which trace is to be found among the rocks of Thurso, were curiously fretted on their external surfaces, and in the other species comparatively smooth. The largest, or coracoidian plate of the ring, as it occurs in the more ornate species (fig. 38), might be readily enough mistaken, when seen with only its surface exposed,

Fig. 38.

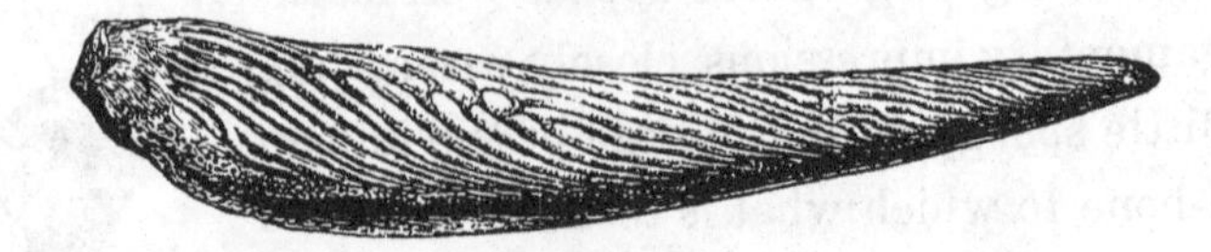

SHOULDER (*i. e.* CORACOID?) PLATE OF ASTEROLEPIS.
(One-third nat. size, linear.)

for the ichthyodorulite of some large fish, allied, mayhap, to the *Gyracanthus formosus* of the Coal Measures; but when detached from the stone, the hollow form and peculiar striæ of the inferior surface serve to establish its true character as a dermal plate. The diagonal furrowings which traversed it, as the twisted flutings traverse a Gothic column moulded after the type of the Apprentice Pillar in Roslin chapel, seem to have underlaid the edge of the opercule; at least I find a similar arrangement in the shoulder-plates of a large species of *Diplopterus*, which are deeply grooved and furrowed where the opercule rested, as if with the design of keeping up a communication between the branchiæ and the external element, even when the gill-cover was pressed closely down upon them. And,—as in these shoulder plates of the *Diplopterus* the furrows yield their place beyond the edge of the opercule to the punctulated enamel common to the outer surface of all the creature's external plates and scales,—we find them yielding their place, in the shoulder-plates of the *Asterolepis*, to the starred tubercles.

A few detached bones, that bear on their outer surfaces the dermal markings, must have belonged to that angular-shaped portion of the head which intervened between the cranial buckler and the intermaxillary bone; but the key for assigning to them their proper place is still to find; and I suspect that no amount of skill on the part of the comparative anatomist will ever qualify him to complete the work of restoration without it. I have submitted to the reader the cranial bucklers of *five* several genera of the ganoids of the Old Red Sandstone ; but no amount of study bestowed on these would enable even the most skilful ichthyologist to restore a *sixth* ; nor is the lateral area of the head, which was, I find, variously occupied in each genus, less difficult to restore than the buckler which surmounted it. Two of the more entire of these dermal bones I have figured (fig. 39, *a* and *b*), in the hope of assisting future inquirers, who,

Fig. 39.

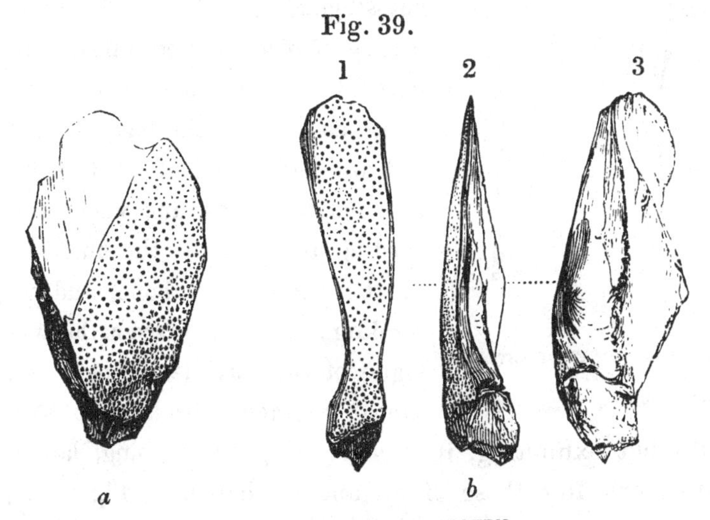

DERMAL BONES OF ASTEROLEPIS.
(One-third nat. size, linear.)

were they to pick up all the other plates, might yet be unable, lacking the figured ones, to complete the whole. The curiously-shaped plate *a*, represented in its various sides by the figures 1, 2, 3, is of an acutely angular form in the transverse section (the external surface, 1, forming an angle which varies from thirty to forty-five degrees with the base, 3) ; and as it lay, it is probable, when in its original place, immediately under the edge of the cranial buckler, it may have served to commence the line of deflection from the flat top of the head to the steep descent of the sides, just as what are technically termed the *spur*-stones in a gable-head serve to commence the line of deflection from the vertical outline of the wall to the inclined line of the roof, or as the spring-stones of an arch serve to commence the curve. A few internal bones in my possession are curious, but exceedingly puzzling. The bone *a*, fig. 40, which resembles a rib, or branchiostegous ray, of one of the ordinary fishes, formed apparently part of that osseous *style* which in fishes such as the haddock and cod we find attached to the suite of shoulder-bones, and which according to Cuvier is the analogue of the coracoidian bone, and according to Professor Owen, the analogue of the clavicle. Fig. *b* is a mere fragment, broken at both ends, but exhibiting, in a state of good keeping, lateral expansions, like those of an ancient halbert. Fig. *c*, 41, which is also a fragment, though a more considerable one, bears in its thicker and straighter edge a groove like that of an ichthyodorulite, which, however, the bone itself in no degree resembles. Fig. *d* is a flat bone, of a type common in the skeleton of fishes, but which in mammals we find exemplified in but the scapulars. It seems, like these, to have furnished the base to which some suite of moveable bones was articulated,—in all likelihood that proportion of the carpal bonelets of the pectoral fins which are attached in

Fig. 40.

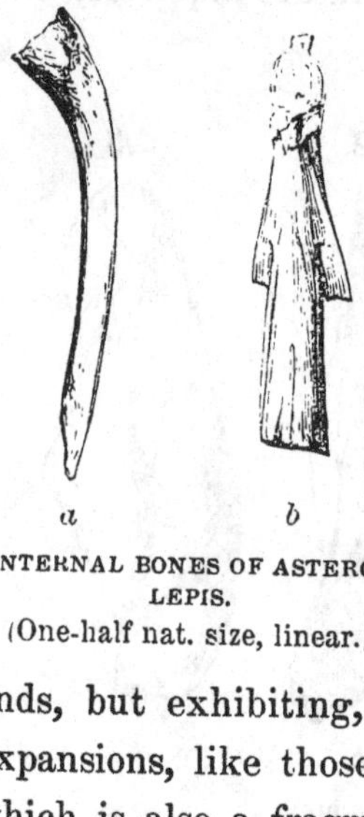

a *b*

INTERNAL BONES OF ASTEROLEPIS.
(One-half nat. size, linear.)

Fig. 41.

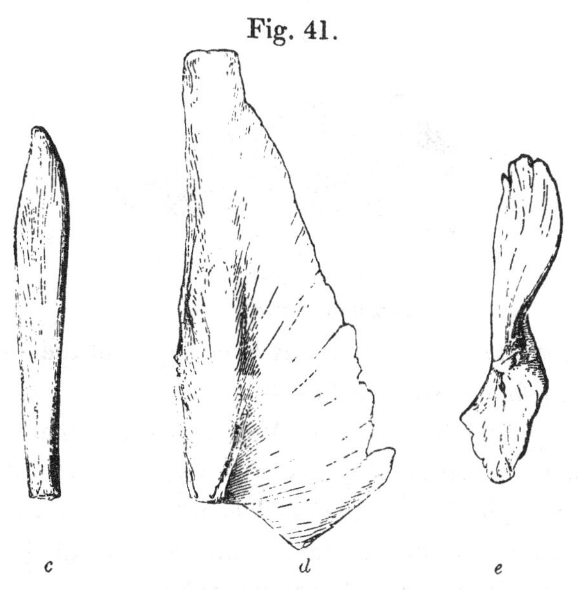

INTERNAL BONES OF ASTEROLEPIS.
(One-third nat. size, linear.)

the osseous fishes to its apparent homologue, the radius. Fig. *e*, a slim light bone, which narrows and thickens in the centre, and flattens and broadens at each end, was probably a scapula or shoulder-blade,—a bone which in most fishes *splices on*, as a sailor would say, by squamose jointings, to the coracoidian bone at the one end, and the super-scapular bone at the other. As indicated by its size, it must have belonged to a small individual : it is, however, twice as long, and about six times as bulky, as the scapula of a large cod.

Of the bone represented in fig. 42, I have determined, from a Cromarty specimen, the place and use : it formed the interior base to which one of the ventral fins was attached. In all fishes the bones of the hinder extremities are inadequately represented : in none do we find the pelvic arch complete ; and to that nether portion of it which we do find represented, and which Professor Owen regards as the homologue

Fig. 42.

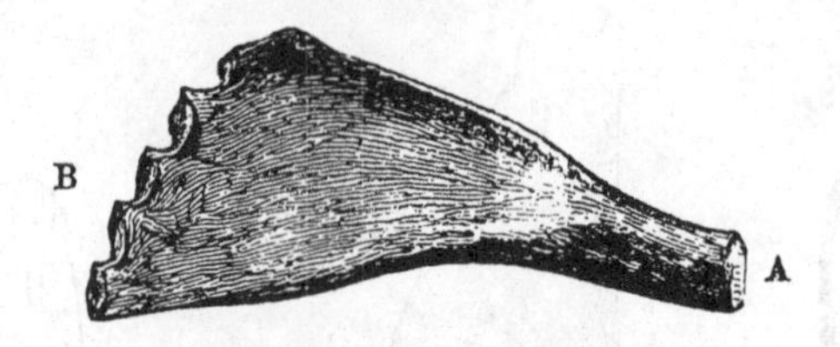

ISCHIUM OF ASTEROLEPIS.
(One-half nat. size, linear.)

of the *os ischium* or hip-bone, the homologues of the metatarsal and toe-bones are attached, to the exclusion of the bones of the thigh and leg. In the Abdominales,—fishes such as the salmon and carp,—that have the ventrals placed behind the abdomen, in the position analogous to that in which the hinder legs of the reptiles and mammals occur, the ischiatic bones generally exist as flat triangular plates, with their heads either turned *inwards* and downwards, as in the herring, or *outwards* and downwards, as in the pike ; whereas in some of the cartilaginous fishes, such as the Rays and Sharks, they exist as an undivided cartilaginous band, stretched transversely from ventral to ventral. And such, with but an upward direction, appears to have been their position in the *Asterolepis*. They seem to have united at the narrow neck A, over the middle of the lower portion of the abdomen ; and to the notches of the flat expansion B,—notches which exactly resemble those of the immensely developed carpal bones of the Ray,—five metatarsal bones were attached, from which the fin expanded. It is interesting to find the number in this ancient representative of the vertebrata restricted to five,—a number greatly exceeded in most of the existing fishes, but which is the true normal number of the vertebrate sub-kingdom, as shown in all the higher examples, such as man, the *quadrumana*, and in most of the *carnaria*. The form of this bone somewhat resembles that of the analogous bone in those fishes, such as the perch and gurnard, cod and haddock, which have their ventrals suspended to the scapular belt; but its position in the Cromarty specimen, and that of the ventrals in the various specimens of the Celacanth family in which their place is still shown, forbids the supposition that *it* was so suspended,—a circumstance in keeping with all the existing geological evidence on the subject, which agrees in indicating, that of the low type of fishes that have, monster-like, their *feet* attached to their necks, the Old Red Sandstone does not afford a trace. This inferior type, now by far the most prevalent in the ichthyic division of the animal kingdom, does not seem to have been introduced until near the close of the Secondary period, long after the fish had been degraded from its primal place in the fore-front of creation. In one of my specimens a few fragments of the rays are preserved (fig. 43, *b*.) They are about the eighth part of an inch in diameter; depressed in some cases in the centre, as if, over the internal hollow formed by the decay of the cartilaginous centre, the bony crust of which they are composed had given way; and, like the rays of the thornback, they are thickened at the joints, and at the processes by which they were attached to the ischiatic base. It may be proper I should here state, that of some of the internal bones figured above I have no better evidence that they belonged to the *Asterolepis*, than that they occur in the same beds with the dermal plates which bear the characteristic star-like markings,—that they are of very considerable size,—and that they formed no part of the known fishes of the formation.

Fig. 43.

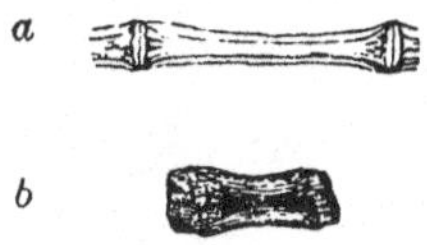

a. *Single joint of ray of Thornback.*
b. *Single joint of ray of Asterolepis.*

On exactly the same grounds I infer, that certain large coprolites of common occurrence in the Thurso flagstones, which contain the broken scales of Dipterians, and exhibit a curiously twisted form (fig. 44), also belonged to the *Asterolepis*; and

Fig. 44.

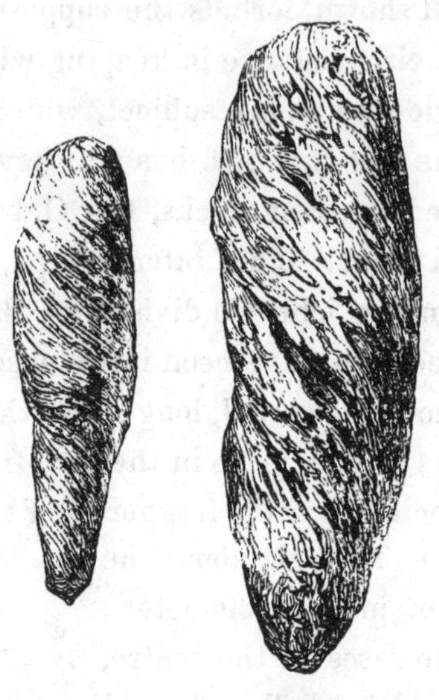

COPROLITES OF ASTEROLEPIS.
(Nat. size.*)

from these, that the creature was carnivorous in its habits,—an inference which the character of its teeth fully corroborates; and farther, that, like the sharks and rays, and some of the extinct enaliosaurs, it possessed the spiral disposition of intestine. Paley, in his chapter on the compensatory contrivances palpable in the structure of various animals, refers to a peculiar substitutory provision which occurs in a

* One of the Thurso coprolites in my possession is about one-fourth longer than the larger of the two specimens figured here and nearly thrice as broad.

certain amphibious animal described in the Memoirs of the French Academy. "The reader will remember," he says, "what we have already observed concerning the *intestinal* canal,—that its length, so many times exceeding that of the body, promotes the extraction of the chyle from the aliment, by giving room for the lacteal vessels to act upon it through a greater space. This long intestine, whenever it occurs, is in other animals disposed in the abdomen from side to side, in returning folds. But in the animal now under our notice, the matter is managed otherwise. The same intention is mechanically effectuated, but by a mechanism of a different kind. The animal of which I speak is an amphibious quadruped, which our authors call the alopecias or sea-fox. The intestine is straight from one end to the other; but in this straight, and consequently short intestine, is a winding, cork-screw, spiral passage, through which the food, not without several circumvolutions, and, in fact, by a long route, is conducted to its exit. Here the shortness of the gut is *compensated* by the obliquity of the perforation." This structure of intestine, which all the true placoids possess, and at least the Sturiones among existing ganoids, seems to have been an exceedingly common one during both the Palæozoic and Secondary periods. It has left its impress on all the better preserved coprolites of the Coal Measures, so abundant in the shales of Newhaven and Burdie House, and on those of the Lias and Chalk. It seems to be equally a characteristic of well nigh all the bulkier coprolites of the Lower Old Red Sandstone.* In these, however, it

* In two of these, in a collection of several score, I have failed to detect the spiral markings, though their state of keeping is decidedly good. There are other appearances which lead me to

manifests a peculiar trait, which I have failed to detect in any of the recent fishes; nor have I yet seen it indicated, in at least the same degree, by the Carboniferous or Secondary coprolitic remains. In the bowels which moulded the coprolites of Lyme-Regis, of the Chalk, and of the Newhaven and Granton beds, a single screw must have winded within the cylindrical tube, as a turnpike stair winds within its hollow shaft; and such also is the arrangement in the existing Sharks and Rays; whereas the bowels which moulded the coprolites of the Lower Old Red Sandstone must have been traversed by triple or quadruple screws laid closely together, as we find the stalk of an old-fashioned wine-glass traversed by its thickly-set spiral lines of thread-like china. And so, while on the surface of both the Secondary and Carboniferous coprolites there is space between the screw-like lines for numerous cross markings that correspond to the thickly set veiny branches which traverse the sides of the recent placoid bowel, the entire surface of the Lower Old Red coprolites is traversed by the spiral markings. Is there nothing strange in the fact, that after the lapse of mayhap millions of years,—nay, it is possible, millions of ages,—we should be thus able to detect at once general resemblance and special dissimilarity in even the most perishable parts of the most ancient of the ganoids?

I must advert, in passing, to a peculiarity exemplified in the state of keeping of the bones of this ancient ganiod, in at least the deposits of Orkney and Caithness. The original animal matter has been converted into a dark-coloured bi-

---

suspect that the *Asterolepis* was not the only large fish of the Lower Old Red Sandstone; but my facts on the subject are too inconclusive to justify aught more than sedulous inquiry.

tumen, which in some places, where the remains lie thick, pervades the crevices of the rocks, and has not unfrequently been mistaken for coal. In its more solid state it can hardly be distinguished, when used in sealing a letter, —a purpose which it serves indifferently well,—from black wax of the ordinary quality; when more fluid, it adheres scarce less strongly to the hands than the coal-tar of our gas-works and dock-yards. Underneath a specimen of *Asterolepis*, first pointed out to me in its bed among the Thurso rocks by Mr Dick, and which, at my request, he afterwards raised and sent me to Edinburgh, packed up in a box, there lay a quantity of thick tar, which stuck as fast to my fingers, on lifting out the pieces of rock, as if I had laid hold of the planking of a newly tarred yawl. What had been once the nerves, muscles, and blood of this ancient ganoid still lay under its bones, and reminded me of the appearance presented by the remains of a poor suicide, whose solitary grave, dug in a sandy bank in the north of Scotland, had been laid open by the encroachments of a river. The skeleton, with pieces of the dress still wrapped round it, lay at length along the section; and, for a full yard beneath, the white dry sand was consolidated into a dark-coloured pitchy mass, by the altered animal matter which had escaped from it percolating downwards, in the process of decay.

In consequence of the curious chemical change which has thus taken place in the animal juices of the *Asterolepis*, its remains often occur in a state of beautiful preservation: the pervading bitumen, greatly more conservative in its effects than the oils and gums of an old Egyptian undertaker, has maintained, in their original integrity, every scale, plate, and bone. They may have been much broken ere they were first committed to the keeping of the rock, or in disentangling them from its rigid embrace ; but they have, we find, caught no harm when under its care. Ere the skeleton of the Bruce, disinterred after the lapse of five centuries, was re-committed to the tomb, such measures were taken to secure its preservation, that, were it to be again disinterred, even after as many more centuries had passed, it might be found retaining unbroken its gigantic proportions. There was molten pitch poured over the bones, in a state of sufficient fluidity to permeate all the pores, and fill up the central hollows, and which, soon hardening around them, formed a bituminous matrix, in which they may lie unchanged for a thousand years. Now, exactly such was the process to which nature resorted with these gigantic skeletons of the Old Red Sandstone. Like the bones of the Bruce, they are bones steeped in pitch ; and so thoroughly is every pore and hollow still occupied, that, when cast into the fire, they flame like torches. Though black as jet, they still retain, too, in a considerable degree, the peculiar *qualities* of the original substance. The late Mr George Sanderson of Edinburgh, one of the most ingenious lapidaries in the kingdom, and a thoroughly intelligent man, made several preparations for me, for microscopic examination, from the teeth and bones ; and though they were by far the oldest vertebrate remains he had ever seen, they exhibited, he informed me, in the working, more of the characteristics of recent teeth and bone than any other fossils he had ever operated upon. Recent bone, when in the course of being reduced on the wheel to the degree of thinness necessary to secure transparency, is apt, under the heat induced by the friction, to acquire a springy elasticity, and to start up from the glass slip to which it has been cemented ; whereas bone in the fossil state usually lies as passive, in such circumstances, as the stone which envelopes it. Mr Sanderson was, however, surprised to find that the bone of the *Asterolepis* still retained its elasticity, and was scarce less liable, when heated, to start from the glass,—a peculiarity through which he at first lost several preparations. I have seen a human bone that had for ages been partially embedded in a mass of adipocere, partially enveloped in the common mould of a churchyard, exhibit two very different styles of keeping. In the adipocere it was as fresh and green as if it had been divested of the integuments only a few weeks previous; whereas the portion which projected into the mould had become brittle and porous, and presented the ordinary appearance of an old churchyard bone. And what the adipocere had done for the human bone in this case, seems to have been done for the bones of the *Asterolepis* by the animal bitumen.

The size of the *Asterolepis* must, in the larger specimens, have been very great. In all those ganoidal fishes of the Old Red Sandstone that had the head covered with osseous plates, we find that the cranial buckler bore a certain definite proportion,—various in the several genera and species,—to the length of the body. The drawing-master still teaches his pupils to regulate the proportions of the human figure by the seven head-lengths which it contains; and perhaps shows them how an otherwise meritorious draftsman,* much employed half an age ago in drawing for the wood-engraver, used to render his figures squat and ungraceful by making them a head too short. Now, those ancient ganoids which possessed a cranial buckler may, we find, be also measured by head-lengths. Thus, in the *Coccosteus decipiens*, the length of the cranial buckler from nape to snout equal-

* The late Mr John Thurston.

led one-fifth the entire length of the creature from snout to tail. The entire length of the *Glyptolepis* was equal to about five one-half times that of its cranial buckler. The *Pterichthys* was formed in nearly the same proportions. The *Diplopterus* was fully seven times the length of its buckler; and the *Osteolepis* from six and a half to seven. In all the cranial bucklers of the *Asterolepis* yet found, the snout is wanting. The very fine specimen figured in page 75 (fig. 28), terminates abruptly at the little plate between the eyes; the specimen figured in page 74 (fig. 27), terminates at the upper line of the eye. The terminal portion which formed the snout is wanting in both, and we thus lack the measure, or *module*, as the architect might say, by which the proportions of the rest of the creature were regulated. We can, however, very nearly approximate to it. A hyoid plate in my collection (fig. 45) is, I find, so exactly proportioned in size to the cranial buckler (fig. 28), that it might have be-

Fig. 45.

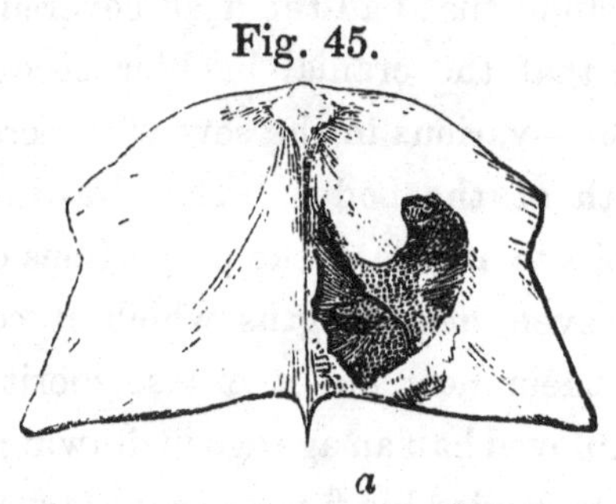

HYOID PLATE OF THURSO ASTEROLEPIS.*
(One-fifth the nat. size, linear.)

* The shaded plate (*a*), accidentally presented in this specimen, belongs to the upper part of the head. It is the posterior frontal plate F, which half encircled the eye-orbit (see fig. 29); and I have introduced it into the print here, as in none of the other prints, or of my other specimens, is its upper surface shown.

longed to the same individual; and by fitting it in its proper place, and then making the necessary allowance for the breadth of the nether jaw, which swept two-thirds around it, and was surmounted by the snout, we ascertain that the buckler, when entire, must have been, as nearly as may be, a foot in length. If the *Asterolepis* was formed in the proportions of the *Coccosteus*, the buckler (fig. 28) must have belonged to an individual five feet in length; if in the proportions of the *Pterichthys* or *Glyptolepis*, to an individual five and a half feet in length; and if in those of the *Diplopterus* or *Osteolepis*, to an individual of from six and a half to seven feet in length. Now I find that the hyoid plate can be inscribed,—such is its form,—in a semicircle, of which the nail-shaped ridge in the middle (if we strike off a minute portion of the sharp point, usually wanting in detached specimens), forms very nearly the radius, and of which the diameter equals the breadth of the cranial buckler, along a line drawn across at a distance from the nape, equal to two-thirds of the distance between the nape and the eyes. Thus, the largest diameter of a hyoid plate which belonged to a cranial buckler a foot in length is, I find, equal to seven one-quarter inches, while the length of its nail somewhat exceeds three five-eighth inches. The nail of the Stromness specimen measures five and a half inches. It must have run along a hyoid plate eleven inches in transverse breadth, and have been associated with a cranial buckler eighteen one-eighth inches in length; and the *Asterolepis* to which it belonged must have measured from snout to tail, if formed, as it probably was, in the proportions of its brother Celacanth the *Glyptolepis*, eight feet three inches; and if in those of the *Diplopterus*, from nine feet nine to ten feet six inches. This oldest of Scottish fish,—this earliest-born of the ganoids yet known,—was at least as bulky as a large porpoise.

It was small, however, compared with specimens of the *Asterolepis* found elsewhere. The hyoid plate figured in page 86 (fig. 36),—a Thurso specimen which I owe to the kindness of Mr Dick,—measures nearly fourteen inches, and the cranial buckler of the same individual, fifteen one-fourth inches, in breadth. The latter, when entire, must have measured twenty-three one-half inches in length; and the fish to which it belonged, if formed in the proportions of the *Glyptolepis*, ten feet six inches; and if in those of the *Diplopterus,* from twelve feet five to thirteen feet eight inches in length. Did the shield still exist in its original state as a buckler of tough, enamel-crusted bone, it might be converted into a Highland target, nearly broad enough to cover the ample chest of a Rob Roy or Allan M'Aulay, and strong enough to dash aside the keenest broadsword. Another hyoid plate found by Mr Dick measures sixteen one-half inches in breadth; and a cast in the British Museum, from one of the Russian specimens of Professor Asmus (fig. 46), twenty-four inches. The individual to which this last plate belonged must, if built in the shorter proportions, have measured eighteen, and if in the longer, twenty-three feet in length. The two hyoid plates of the specimen of *Holoptychius* in the British Museum measure but four and a half inches along that transverse line in which the Russian *Asterolepis* measures two feet, and the largest Thurso specimen sixteen inches and a half. The maxillary bone of a cod-fish two and a half feet from snout to tail measures three inches in length. One of the Russian maxillary bones in the possession of Professor Asmus measures in length twenty-eight inches. And that space circumscribed by the

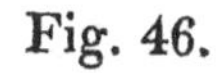

Fig. 46.

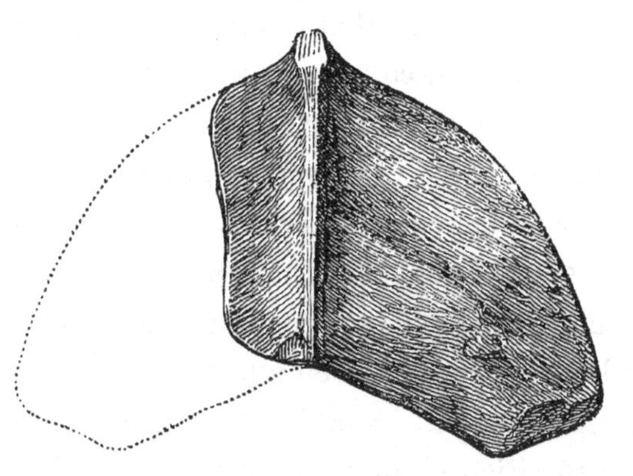

HYOID PLATE OF RUSSIAN ASTEROLEPIS.

(One-twelfth the natural size, linear.)

sweep of the lower jaw which it took, in the Russian specimen, a hyoid plate twenty-four inches in breadth to fill, could be filled in the two-and-a-half-feet cod by a plate whose breadth equalled but an inch and a half. Thus, in the not unimportant circumstance of size, the most ancient ganoids yet known, instead of taking their places, agreeably to the demands of the development hypothesis, among the sprats, sticklebacks, and minnows of their class, took their place among its huge basking sharks, gigantic sturgeons, and bulky sword-fishes. They were giants, not dwarfs.

But what of their organization? Were they fishes low or high in the scale? On this head we can, of course, determine merely by the analogies which their structure exhibits to that of fishes of the existing period; and these point in three several directions;—in two of the number, directly on genera of the high ganoid order; and in the third, on the still higher placoids and enaliosaurs. No trace of vertebræ has yet been found; and so we infer,—lodging, however, a precautionary protest, as the evidence is purely negative, and therefore in some degree inconclusive,—that the vertebral column of the *Asterolepis* was, like that of the sturgeon, cartilaginous. Respecting its external covering, we positively know, as has been already shown, that, like the *Lepidosteus* of America and the *Polypterus* of the Nile, it was composed of strong plates and scales of solid bone ; and, regarding its dentition, that, as in these last genera, and even more decidedly than in these, it was of the mixed ichthyic-reptilian character,—an outer row of thickly-set fish-teeth being backed by an inner row of thinly-set reptile-teeth. And its form of coprolite indicates the spiral disposition of intestine common to the Rays and Sharks of the existing period, and of the Ichthyosauri of the Secondary ages. Instead of being, as the development hypothesis would require, a fish low in its organization, it seems to have ranged on the level of the highest ichthyic-reptilian families ever called into existence. Had an intelligent being, ignorant of what was going on upon earth during the week of creation, visited Eden on the morning of the sixth day, he would have found in it many of the inferior animals, but no trace of man. Had he returned again in the evening, he would have seen, installed in the office of keepers of the garden, and ruling with no tyrant sway as the humble monarchs of its brute inhabitants, two mature human creatures, perfect in their organization, and arrived at the full stature of their race. The entire evidence regarding them, in the absence of all such information as that imparted to Adam by Milton's angel, would amount simply to this, that in the morning man *was not*, and that in the evening he *was*. There, of course, could not exist, in the circumstances, a single appearance to sanction the belief that the two human creatures whom he saw walking together among the trees at sunset had been "developed from infusorial points," not created mature.

The evidence would, on the contrary, lie all the other way. And in no degree does the geologic testimony respecting the earliest ganoids differ from what, in the supposed case, would be the testimony of Eden regarding the earliest men. Up to a certain point in the geologic scale we find that the ganoids *are not;* and when they at length make their appearance upon the stage, they enter large in their stature and high in their organization.

FISHES OF THE SILURIAN ROCKS—UPPER AND LOWER.

THEIR RECENT HISTORY, ORDER, AND SIZE.

BUT the system of the Old Red Sandstone represents the *second*, not the *first*, great period of the world's history. There was a preceding period at least equally extended, perhaps greatly more so, represented by the Upper and Lower Silurian formations. And what is the testimony of this morning period of organic existence, in which, so far as can yet be shown, vitality, in the planet which man inhabits, and of whose history or productions he knows anything, was first associated with matter? May not the development hypothesis find a standing in the system representative of this earliest age of creation, which it fails to find in the system of the Old Red Sandstone?

It has been confidently asserted, not merely that it *may*, but that it *does*. Ever since the publication, in 1839, of Sir Roderick Murchison's great work on the Silurian System, it had been known that the remains of fishes occur in a bed of the "Ludlow Rock,"—one of the most modern deposits of the *Upper* Silurian division; and subsequent discoveries, both in England and America, had shown that even the *base* of this division has its ichthyic organisms. But for year after year, the lower half of the system,—a division more than three thousand feet in thickness,—had failed, though there were hands and eyes busy among its deposits, to yield any vertebrate remains. During the earlier half of the first great period of organic existence, though the polyparia, radiata, articulata, and mollusca, existed, as their remains testified, by myriads, fish had, it was held, not yet entered upon the scene ; and the assertors of the development theory founded largely on the presumed fact of their absence. " It is still customary," says the author of the " Vestiges of Creation," in his volume of " Explanations," " to speak of the earliest fauna as one of an elevated kind. When rigidly examined, it is not found to be so. IN THE FIRST PLACE, IT CONTAINS NO FISH. There were seas supporting crustacean and molluscan life, but *utterly devoid of a class of tenants who seem able to live in every example of that element which supports meaner creatures.* This single fact, that only invertebrated animals now lived, is surely in itself a strong proof that, in the course of nature, *time* was necessary for the creation of the superior creatures. And if so, it undoubtedly is a powerful evidence of such a theory of development as that which I have presented. If not, let me hear an equally plausible reason for the great and amazing fact, that seas were for numberless ages destitute of fish. I fix my opponents down to the consideration of this fact, so that no diversion respecting high molluscs shall avail them." And how is this bold challenge to be met ?

Most directly, and after a fashion that at once discomfits the challenger.

It might be rationally enough argued in the case, that the author of the " Vestiges" was building greatly more on a piece of purely negative evidence,—the presumed absence of fish from the Lower Silurian formations,—than purely negative evidence is, from its nature as such, suited to bear; that only a very few years had passed since it was known that vertebrate remains occurred in the *Upper* Silurian, and only a few more since they had been detected in the Old Red Sandstone; nay, that within the present century their frequent occurrence in even the Coal Measures was scarce suspected; and that, as his argument, had it been founded twelve years ago on the supposed absence of fishes from the Upper Silurian, or twenty years ago on the supposed absence of fishes from the Old Red Sandstone, would have been quite as plausible in reference to its negative data then as in reference to its negative data now, so it might now be quite as erroneous as it assuredly would have been then. Or it might be urged, that the fact of the absence of fish from the Lower Silurians, even were it really a fact, would be in no degree less reconcileable with the theory of creation by direct act, than with the hypothesis of gradual development. The fact that Adam did not exist during the first, second, third, fourth, and fifth days of the introductory week of Scripture narrative, furnishes no argument whatever against the fact of his *creation* on the sixth day. And the remark would of course equally apply to the non-existence of fishes during the Lower Silurian period, had they been really non-existent at the time, and to their sudden appearance in that of the Upper. But the objection admits of a greatly more conclusive answer. "I fix my opponents down," says the author of the "Vestiges," "to the consideration of this fact," *i. e.* that of the absence of fishes from the earliest fossiliferous formations. And I, in turn, fix you down, I reply, to the consideration of the antagonist fact, not negative, but positive, and now, in the course of geological discovery, fully established, that fishes were *not* absent from the earliest fossiliferous formations. From none of the great geological formations were fishes absent,—not even from the formations of the Cambrian division. "The Lower Silurian," says Sir Roderick Murchison, in a communication with which, in 1847, he honoured the writer of these chapters, "is no longer to be viewed as an invertebrate period; for the *Onchus* (species not yet decided) has been found in the Llandeilo Flags and in the Lower Silurian rocks of Bala. In one respect I am gratified by the discovery; for the form is so very like that of the *Onchus Murchisoni* of the Upper Ludlow rock, that it is clear the Silurian system is one great natural-history series, as is proved, indeed, by all its other organic remains." It may be mentioned further, in addition to this interesting statement, that the Bala spine was detected in its calcareous matrix by the geologists of the Government Survey, and described to Sir Roderick as that of an *Onchus*, by a very competent authority in such matters,—Professor Edward Forbes; and that the annunciation of the existence of spines of fishes in the Llandeilo Flags we owe to one of the most cautious and practised geologists of the present age,—Professor Sedgwick of Cambridge.

So much for the *fact* of the existence of vertebrata in the Lower Silurian formations, and the *argument* founded on their presumed absence. Let me now refer,—their presence being determined,—to the tests of size and organization. Were these Silurian fishes of a bulk so inconsiderable as in any degree to sanction the belief that they had been developed shortly before from microscopic points? Or were they of a structure so low as to render it probable that their development was at the time incomplete? Were they, in other words, the embryos and fœtuses of their class? or did they, on the contrary, rank with the higher and larger fishes of the present time?

It is of importance that not only the direct *bearing*, but also the actual *amount*, of the evidence in this case, should be fairly stated. So far as it extends, the testimony is clear; but it does not extend far. All the vertebrate remains yet detected in the Silurian System, if we except the debris of the Upper Ludlow bone-bed, might be sent through the Post-Office in a box scarcely twice the size of a copy of the "Vestiges." The naturalist of an exploring party, who, in crossing some unknown lake, had looked down over the side of his canoe, and seen a few fish gliding through the obscure depths of the water, would be but indifferently qualified, from what he had witnessed, to write a history of *all* its fish. Nor, were the some six or eight individuals of which he had caught a glimpse to be of small size, would it be legitimate for him to infer that only small-sized fish lived in the lake; though, were there to be some two or three large ones among them, he might safely affirm the contrary. Now, the evidence regarding the fishes of the Silurian formation very much resembles what that of the naturalist would be, in the supposed case, regarding the fishes of the unexplored lake; with, however, this difference, that as the deposits of the ancient system in which they occur have been examined for years in various parts of the world, and all its characteristic organisms, save the ichthyic ones, found in great abundance and fine keeping, we may conclude that the fish of the period were comparatively few. The palæontologist, so far as the question of number is involved, is in the circumstances, not of the naturalist who has only once crossed the unknown lake, but of the angler who, day after day, casts his line into some inland sea abounding in shell-fish and crustacea, and, after the lapse of months, can scarce detect a nibble, and, after the lapse of years, can reckon up all the fish which he has caught as considerably under a score. The existence of this great division of the animal kingdom, like that of the earlier reptiles during the Carboniferous period, did not form a prominent characteristic of those ages of the earth's history in which they began to be.

The earliest discovered vertebral remains of the system,—those of the Upper Ludlow rock,—were found in digging the foundations of a house at Ludford, on the confines of Shropshire, and submitted, in 1838, by Sir Roderick Murchison to Agassiz, through the late Dr Malcolmson of Madras. I used at the time to correspond on geological subjects with Dr Malcolmson,—an accomplished geologist and a good man, too early lost to science and his friends,—and still remember the interest which attached on this occasion to his communication bearing the Paris post-mark, from which I learned for the first time that there existed ichthyic fragments greatly older than even the ichthyolites of the Lower Old Red Sandstone, and which made me acquainted with Agassiz's earliest formed decision regarding them. Though existing in an exceedingly fragmentary condition,—for the materials of the thin dark-coloured layer in which they had lain seemed as if they had been triturated in a mortar,—the ichthyologist succeeded in erecting them into six genera; though it may be very possible,—as some of these were formed for the reception of detached spines, and others for the reception of detached teeth,—that, as in the case of *Dipterus* and *Asterolepis*, the fragments of but a single genus may have been multiplied into two genera or more. And minute scale-like markings, which mingled with the general mass, and were at first regarded as the impressions of real scales, have been since recognised as of the same character with the scale-like markings of the *Seraphim* of Forfarshire, a huge crustacean. Even admitting, however, that a set of teeth and spines, with perhaps the shagreen points represented in page 30, fig. 2, *b*, in addition, may have all belonged to but a single species of fish, there seem to be materials enough among the remains found, for the erection of two species more. And we have evidence that at least two of the three kinds were fishes of the Placoid order (*Onchus Murchisoni* and *Onchus tenuistriatus*), and,—as the supposed scales must be given up,—no good evidence that the other kind was not. The ichthyic remains of the Silurian System next discovered were first introduced to the notice of geologists by Professor Phillips, at the meeting of the British Association in 1842.* They occurred, he stated, in a quarry near Hales End, at the base of the Upper Ludlow rock, im-

---

* "Mr Phillips proceeded to describe some remains of a small fish, resembling the *Cheiracanthus* of the Old Red Sandstone, scales and spines of which he had found in a quarry at Hales End, on the western side of the Malverns. The section presented beds of the Old Red Sandstone inclined to the west; beneath these were arenaceous beds of a lighter colour, forming the junction with Silurian shales; these, again, passing on to calcareous beds in the lower part of the quarry, containing the corals and shells of the Aymestry Limestone, of their agreement with which stronger evidence might be obtained elsewhere. He had found none of these scales in the junction beds or in the Upper Ludlow shales; but about sixty or one hundred feet lower, just above the Aymestry Limestone, his attention had been attracted to discoloured spots on the *surface* of the beds, which, upon microscopic examination, proved to be the minute scales and spines before mentioned. These remains were only apparent on the surface, whilst the 'fish-bed' of the Upper Ludlow rock, as it usually occurred, was an inch thick, consisting of innumerable small teeth and spines."—*Report, in* "Athenæum" *for* 1842, *of the Proceedings of the Twelfth Meeting of British Association (Manchester).*

mediately over the Aymestry Limestone, and were so exceedingly diminutive, that they appeared to the naked eye as mere discoloured spots; but resolved under the microscope into scattered groupes of minute spines, like those of the *Cheiracanthus*, with what seemed to be still more minute *scales*, or perhaps,—what in such circumstances could scarce be distinguished from scales,—shagreen points of the scale-like type. The next ichthyic organism detected in the Silurian rocks occurred in the Wenlock Limestone, a considerably lower and older deposit, and was first described in the "Edinburgh Review" for 1845 by a vigorous writer and masterly geologist (generally understood to be Professor Sedgwick of Cambridge), as "a characteristic portion of a fish undoubtedly belonging to the Cestraciont family of the placoid order." In the "American Journal of Science" for 1846, Professor Silliman figured, from a work of the States' Surveyors, the defensive spine of a placoid found in the Onondago Limestone of New York,—a rock which occurs near the base of the Upper Silurian System, as developed in the western world;* and in the same passage he made reference to a mutilated spine detected in a still lower American deposit,—the Oriskany Sandstone. In the "Geological Journal for 1847, it was announced by Professor Sedgwick,

* "This is the lowest position" (that of the Onondago Limestone) "in the State of New York in which any remains have been found higher in the scale of organized beings than *Crustacea*, with the exception of an imperfectly preserved fish-bone discovered by Hall in the Oriskany Sandstone. That specimen, together with the defensive fish-bone found in this part of the New York system, furnishes evidences of the existence of animals belonging to the class *vertebrata* during the deposition of the middle part of the protozoic strata."—*American Journal of Science and Arts for* 1846, p. 63.

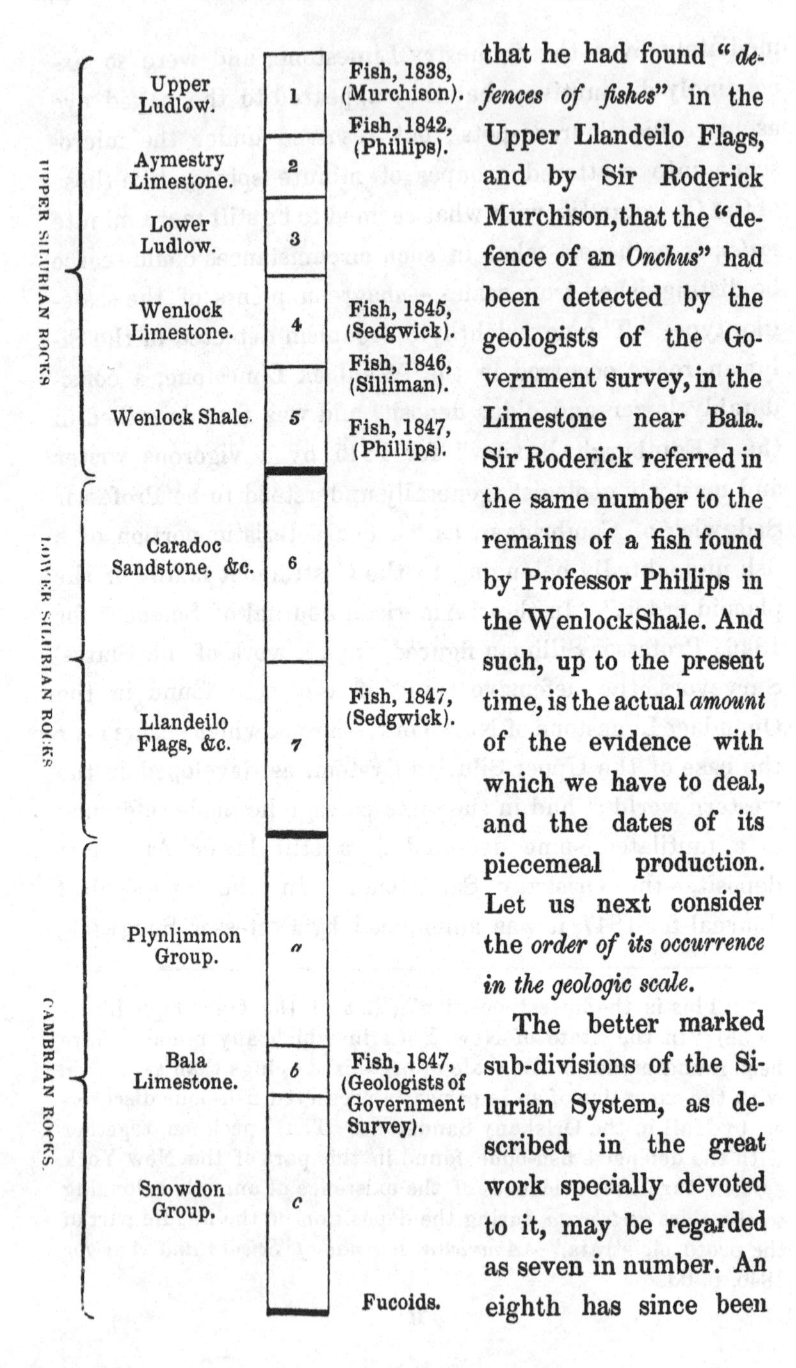

that he had found "*defences of fishes*" in the Upper Llandeilo Flags, and by Sir Roderick Murchison, that the "defence of an *Onchus*" had been detected by the geologists of the Government survey, in the Limestone near Bala. Sir Roderick referred in the same number to the remains of a fish found by Professor Phillips in the Wenlock Shale. And such, up to the present time, is the actual *amount* of the evidence with which we have to deal, and the dates of its piecemeal production. Let us next consider the *order of its occurrence in the geologic scale.*

The better marked sub-divisions of the Silurian System, as described in the great work specially devoted to it, may be regarded as seven in number. An eighth has since been added, by the transference of the Tilestones from the lower part of the Old Red Sandstone group, to the upper part of the Silurian group underneath ; but in order the better to show how ichthyic discovery has in its slow course penetrated into the depths, I shall retain the divisions recognised as those of the system when that course began. The highest or most modern Silurian deposit, then (No. 1 of the accompanying diagram), is the *Upper* Ludlow Rock ; and it is in the superior strata of this division that the bone-bed discovered in 1838 occurs ; while the exceedingly minute vertebrate remains described by Professor Phillips in 1842 occur in its base. The division next in the descending order is the Aymestry Limestone (No. 2) ; the next (No. 3) the *Lower* Ludlow rock ; then (No. 4) the Wenlock or Dudley Limestone occurs ; and then, last and oldest deposit of the *Upper* Silurian formation, the Wenlock shale (No. 5). It is in the fourth, or Wenlock Limestone division, that the defensive spine described in the "Edinburgh Review" for 1845 as the oldest vertebrate organism known at the time, was found ;* while the vertebrate organism found by Professor Phillips belongs to the fifth, or base deposit of the Upper Silurian. Further, the American spines of Onondago and Oriskany, described in 1846, occurred in rocks deemed contemporary with those of the Wenlock division. We next cross the line which separates the base of the Upper from the top of the Lower Silurian deposits, and find a great arenaceous formation (No. 6), known as the Caradoc Sandstones ; while the Llandeilo Flags (No. 7), the formation upon which the sandstones rest, compose, according to the sections of Sir Roderick, published in 1839, the lowest de-

* "The shales *alternating* with the Wenlock Limestone." (*Edinburgh Review.*)

posit of the Lower Silurian rocks. And it is in the upper part of this lowest member of the system that the ichthyic defences announced in 1847 by Professor Sedgwick occur. Vertebrate remains have now been detected in the same relative position in the *seventh* and *most ancient* member of the system, that they were found to occupy in its *first* and *most modern* member ten years ago. But this is not all. Beneath the Lower Silurian division there occur vast fossiliferous deposits, to which the name "Cambrian System" was given, merely provisionally, by Sir Roderick, but which Professor Sedgwick still retains as representative of a distinct geologic period; and it is in these, greatly below the Lower Silurian base line, as drawn in 1839, that the Bala Limestones occur. The Plynlimmon rocks (*a*),—a series of conglomerate, grauwacke, and slate beds, several thousand yards in thickness,—intervene between the Llandeilo Flags and the Limestones of Bala (*b*). And, of consequence, the defensive spine of the *Onchus*, announced in 1847 as detected in these limestones by the geologists of the Government Survey, must have formed part of a fish that perished many ages ere the oldest of the Lower Silurian formations *began* to be deposited.

Let us now, after this survey of both the amount of our materials, and the order and time of their occurrence, pass on to the question of size, as already stated. Did the ichthyic remains of the Silurian System hitherto examined and described belong to large or to small fishes? The question cannot be altogether so conclusively answered as in the case of those ganoids of the Lower Old Red Sandstone whose dermal skeletons indicate their original dimensions and form. In fishes of the placoid order, such as the Sharks and Rays, the dermal skeleton is greatly less continuous and persistent than in such ganoids as the Dipterians and Celacanths ; and when their remains occur in the fossil state, we can reason, in most instances, regarding the bulk of the individuals of which they formed part, merely from that of detached teeth or spines, whose proportion to the entire size of the animals that bore them cannot be strictly determined. We can, indeed, do little more than infer, that though a large placoid may have been armed with but small spines or teeth, a small placoid could not have borne very large ones. And to this placoid order all the Silurian fish, from the Aymestry Limestone to the Cambrian deposits of Bala inclusive, unequivocally belong. Nor, as has been already said, is there sufficient evidence to show that any of the ichthyic remains of the Upper Ludlow rocks do *not* belong to it. It is peculiarly the order of the system. The Ludlow bone-bed contains not only defensive spines, but also teeth, fragments of jaws, and shagreen points ; whereas, in all the inferior deposits which yield any trace of the vertebrata, the remains are those of defensive spines exclusively. Let us, then, take the defensive spine as the part on which to found our comparison.

One of the best-marked placoids of the Upper Ludlow bone-bed is that *Onchus Murchisoni* to which the distinguished geologist whose name it bears refers, in his communication, as so nearly resembling the oldest placoid yet known,—that of the Bala Limestone. And the living fishes with which the *Onchus Murchisoni* must be compared, says Agassiz, though "the affinity," he adds, "may be rather distant," are those of the genera "*Cestracion, Centrina,* and *Spinax.*" I have placed before me a specimen of recent *Spinax*, of a species well known to all my readers on the sea-coast, the *Spinax Acanthias*, or common dog-fish,—so little a favourite with our fishermen. It measures exactly two feet three inches in length; and of the defensive spines of its two dorsals,—those spear-like thorns on the creature's back immediately in advance of the fins, which so frequently wound the fisher's hand,—the anterior and smaller measures, from base to point, an inch and a half, and the posterior and larger, two inches. I have also placed before me a specimen of *Cestracion Phillippi* (the Port Jackson Shark), a fish now recognised as the truest existing analogue of the Silurian placoids. It measures twenty-two three-fourth inches in length, and is furnished, like *Spinax*, with two dorsal spines, of which the anterior and larger measures from base to point one one-half inch, and the posterior and smaller, one one-fifth inch. But the defensive spine of the *Onchus Murchisoni*, as exhibited in one of the Ludlow specimens, measures, though mutilated at both ends, three inches and five-eighth parts in length. Even though existing but as a fragment, it is as such nearly twice the length of the largest spine of the dog-fish, unmutilated and entire, and considerably *more* than twice the length of the largest spine of the Port Jackson Shark. The spines detected by Professor Phillips, in an inferior stratum of the same upper deposit, were, as has been shown, of microscopic minuteness; and when they seemed to rest on the extreme horizon of ichthyic existence as the most ancient remains of their kind, the author of the "Vestiges" availed himself of the fact. He regarded the little creatures to which they had belonged as the fœtal embryos of their class, or,—to employ the language of the Edinburgh Reviewer,—as "the tokens of Nature's first and half-abortive efforts to make fish out of the lower animals." From the latter editions of his work, the paragraph to which the Reviewer refers has, I find, been expunged; for the horizon has greatly extended, and what seemed to be its line of extreme distance has travelled into the middle of the prospect. But that the passage should have at all existed is a not uninstructive circumstance, and shows how unsafe it is, in more than external nature, to regard the line at which, for the time, the landscape closes, and heaven and earth seem to meet, as in reality the world's end. The Wenlock spine, though certainly not microscopic, is, I am informed by Sir Philip Egerton, of but small size ; whereas the contemporary spine of the Onondago Limestone, though comparatively more a fragment than the spine of the Upper Ludlow *Onchus*,—for it measures only three inches in length,—is at least five times as bulky as the largest spine of

Fig. 47.

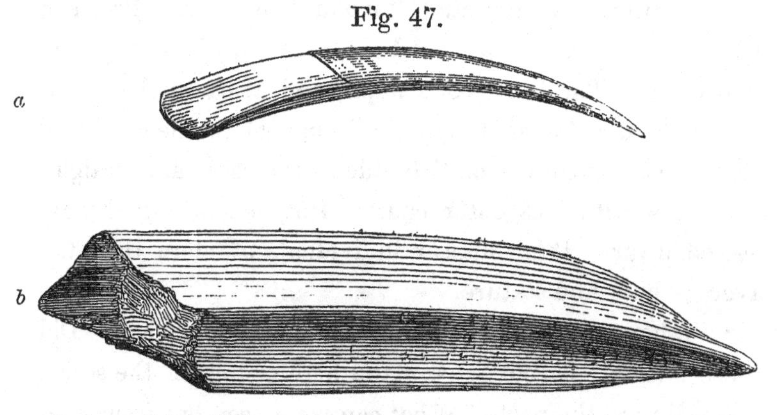

a. *Posterior Spine of Spinax Acanthias.* b. *Fragment of Onondago Spine.*
(Natural Size.)

Spinax Acanthias. Representing one of the massier fishes disporting amid the some four or five small ones, of which, in my illustration, the naturalist catches a glimpse in fording the unknown lake, it at least serves to show that all the Silurian ichthyolites must not be described as small, seeing that not only might many of its undetected fish have been large, but that some of those which *have* been detected were actually so. Another American spine, of nearly the same formation,—for it occurs in a limestone, varying from twenty to seventy feet in thickness, which immediately overlies that of the Onondago deposit, though still more fragmentary than the first, for its length is only two three-eighth inches,—maintains throughout a nearly equal thickness,—a circumstance in itself indicative of considerable size; and in positive bulk it almost rivals the Onondago one. Of the Lower Silurian and Bala fishes no descriptions or figures have yet appeared. And such, up to the present time, is the testimony derived from this department of Geology, so far as I have been able to determine it, regarding the size of the ancient Silurian vertebrata. "No organism," says Professor Oken, "is, nor ever has one been, created, which is not microscopic." The Professor's pupils and abettors, the assertors of the development hypothesis, appeal to the geological evidence as altogether on *their* side in the case; and straightway a few witnesses enter court. But, lo! among the expected dwarfs, there appear individuals of more than the average bulk and stature.

Still, however, the question of organization remains. Did these ancient placoid fishes stand high or low in the scale? According to the poet, "What can we reason but from what we know?" We are acquainted with the placoid fishes of the present time; and from these only, taking analogy as our guide, can we form any judgment regarding the rank and standing of their predecessors, the placoids of the geologic periods. But the consideration of this question, as it is specially one on which the later assertors of the development hypothesis concentrate themselves, I must, to secure the space necessary for its discussion, defer till my next chapter. Meanwhile, I am conscious I owe an apology to the reader for what he must deem tedious minuteness of description, and a too prolix amplitude of statement. It is only by representing things as they actually are, and in the true order of their occurrence, that the effect of the partially selected facts and exaggerated descriptions of the Lamarckian can be adequately met. True, the disadvantages of the more sober mode are unavoidably great. He who feels himself at liberty to arrange his collected shells, corals, and fish-bones, into artistically designed figures, and to select only the pretty ones, will be of course able to make of them a much finer show than he who is necessitated to represent them in the order and numerical proportions in which they occur on some pebbly beach washed by the sea. And such is the advantage, in a literary point of view, of the ingenious theorist, who, in making figures of his geological facts, takes no more of them than suits his purpose, over the man who has to communicate the facts as he finds them. But the homelier mode is the true one. "Could we obtain," says a distinguished metaphysician, "a distinct and full history of all that has passed in the mind of a child, from the beginning of life and sensation till it grows up to the use of reason,—how its infant faculties began to work, and how they brought forth and ripened all the various notions, opinions, and sentiments, which we find in ourselves when we come to be capable of reflection,—this would be a treasure of natural history which would probably give more light into the human faculties than all the systems of philosophers about them since the beginning of the world. But it is in vain," he adds, "to wish for what nature has not put within the reach of our power." In like manner, could we obtain, it may be remarked, a full and distinct account of a single class of the animal kingdom, from its first appearance till the present time, "this would be a treasure of natural history which would cast more light" on the origin of living existences, and the true economy of creation, than all the theories of all the philosophers "since the beginning of the world." And in order to approximate to such a history as nearly as possible,—and it does seem possible to approximate near enough to substantiate the true readings of the volume, and to correct the false ones,—it is necessary that the real vestiges of creation should be carefully investigated, and their order of succession ascertained.

HIGH STANDING OF THE PLACOIDS.—OBJECTIONS CONSIDERED.

We have seen that some of the Silurian placoids were large of size: the question still remains, Were they high in intelligence and organization?

The Edinburgh Reviewer, in contending with the author of the "Vestiges," replies in the affirmative, by claiming for them the first place among fishes. "Taking into account," he says, "the brain and the whole nervous, circulating, and generative systems, they stand at the highest point of a natural ascending scale." They are fishes, he again remarks, that rank among "the very highest types of their class."

"The fishes of this early age, and of all other ages previous to the Chalk," says his antagonist, in reply, "are, for the most part, cartilaginous. The cartilaginous fishes,—*Chondropterygii* of Cuvier,—are placed by that naturalist as a second series in his descending scale; being, however, he says, 'in some measure *parallel to the first.*' How far this is different from their being the highest types of the fish class, need not be largely insisted upon. Linnæus, again, was so impressed by the low characters of many of this order, that he actually ranked them with worms. Some of the cartilaginous fishes, nevertheless, have certain peculiar features of organization, chiefly connected with re-production, in which they excel other fish ; but such features are partly partaken of by families in inferior sub-kingdoms, showing that they cannot truly be regarded as marks of grade in their own class. When we look to the great fundamental characters, particularly to the frame-work for the attachment of the muscles, what do we find ?—why, that of these placoids,—'the highest types of their class,'—it is barely possible to establish their being vertebrata at all, the back-bone having generally been too slight for preservation, although the vertebral columns of later fossil fishes are as entire as those of any other animals. In many of them traces can be observed of the muscles having been attached to the external plates, strikingly indicating their low grade as vertebrate animals. The Edinburgh Reviewer 'highest types of their class' are in reality a separate series of that class, generally inferior, taking the leading features of organization of structure as a criterion, but when details of organization are regarded, stretching farther, both downward and upward, than the other series ; so that, looking at one extremity, we are as much entitled to call them the lowest, as the Reviewer, looking at another extremity, is to call them the 'highest of their class.' Of the general inferiority there can be no room for doubt. Their cartilaginous structure is in the first place analogous to the embryotic state of vertebrated animals in general. The maxillary and intermaxillary bones are in them rudimental. Their tails are finned on the under side only,—an admitted feature of the salmon in an embryotic stage ; and the mouth is placed on the under side of the head,—also a mean and embryotic feature of structure. These characters are essential and important, whatever the Edinburgh Reviewer may say to the contrary : they are the characters which, above all, I am chiefly concerned in looking to, for they are features of embryotic progress, and embryotic progress is the grand key to the theory of development."

Such is the ingenious piece of special pleading which this most popular of the Lamarckians directs against the standing and organization of the earlier fishes. Let us examine it somewhat in detail, and see whether the slight admixture of truth which it contains serves to do aught more than to render current, like the gilding of a counterfeit guinea spread over the base metal, the amount of error which lies beneath. I know not a better example than that which it furnishes, of the entanglement and perplexity which the meshes of an artificial classification, when converted, in argumentative processes, into symbols and abstractions, are sure to involve subjects simple enough in themselves.

Fishes, according to the classification of a preponderating majority of the ichthyologists that have flourished from the earliest times down to those of Agassiz, have been divided into two great series, the *Ordinary* or osseous, and the *Chondropterygii* or cartilaginous. And these two divisions of the class, instead of being ranged consecutively in a continuous line, the one in advance of the other, have been ranged in two parallel lines, the one directly abreast of the other. There is this further peculiarity in the arrangement, that the line of the cartilaginous series, from the circumstance that some of its families rise higher and some sink lower in the scale than any of the ordinary fishes, outflanks the array of the osseous series at both ends. The front which it presents contains fewer genera and species than that of the osseous division ; but, like the front of an army drawn out in single file, it extends along a greater length of ground. And to this long-fronted series of the cartilaginous, or, according to Cuvier, *chondropterygian* fishes, the placoid families of Agassiz belong,—among the rest, the placoids of the Silurian formations, Upper and Lower. But though all the placoids of this latter naturalist be cartilaginous fishes, all cartilaginous fishes are not placoids. The *Sturionidæ* are cartilaginous, and are, as such, ranked by Cuvier among the *Chondropterygii*, whereas Agassiz places them in his ganoid order. Many of the extinct fishes, too, such as the *Acanthodei*, *Dipteridæ*, *Cephalaspidæ*, were, as we have seen, cartilaginous in their internal framework, and yet true ganoids notwithstanding. The principle of Agassiz's classification wholly differs from that of Cuvier and the older ichthyologists; for it is a classification founded, not on the character of the internal, but on that of the cuticular or dermal skeleton. And while to the geologist it possesses great and obvious advantages over every other,—for of the earlier fishes very little more than the cuticular skeleton survives,—it has this farther recommendation to the naturalist, that (in so far at least as its author has been true to his own principles), instead of anomalously uniting the highest and lowest specimens of their class,—the fishes that most nearly approximate to the reptiles on the one hand, and the fishes that sink farthest towards the worms on the other,—it gathers into one consistent order all the individuals of the higher type, distinguished above their fellows by their development of brain, the extensive range of their instincts, and the perfection of their generative systems. Further, the history of animal existences, as recorded in the sedimentary rocks of our planet, reads a recommendation of this scheme of classification which it extends to no other. We find that in the progress of creation the fishes *began to be* by groupes and septs, arranged according to the principle on which it erects its orders. The placoids came first, the ganoids succeeded them, and the ctenoids and cycloids brought up the rear. The march has been marshalled according to an appointed programme, the order of which it is peculiarly the merit of Agassiz to have ascertained.

Now, may I request the reader to mark, in the first place, that what we have specially to deal with at the present stage of the argument are the placoid fishes of the Silurian formations, Upper and Lower. May I ask him to take note, in the second, that the long-fronted *chondropterygian* series of Cuvier, though it includes, as has already been said, the placoid order of Agassiz,—just as the red-blooded division of animals includes the bimana and quadrumana,—is no more to be regarded as *identical* with the placoids, than the red-blooded animals are to be regarded as identical with the apes or with the human family. It simply includes them in the character of *one* of the three great divisions into which it has been separated,—the division ranged, if I may so express myself, on the extreme right of the line ; its middle portion, or main body, being composed of the *Sturiones*, a family on the general level of the osseous fishes ; while, ranged on the extreme left, we find the low division of the *Suctorii, i. e.* Cyclostomi, or Lampreys. But with the middle and lower divisions we have at present nothing to do ; for of neither of them, whether *Sturiones* or *Suctorii*, does the Silurian System exhibit a trace. Further be it remarked, that the scheme of classification which gives an abstract standing to the *Chondropterygii*, is in itself merely a certain perception of resemblance which existed in certain minds, having *cartilage* for its general idea; just as another certain perception of resemblance in one other certain mind had *cuticular skeleton* for its general idea, and as yet another perception of resemblance in yet other certain minds had *red blood* for its general idea. As shown by the disparities which obtain among the section which the scheme serves to separate from the others, it no more determines rank or standing than that greatly more ancient scheme of classification into "ring-streaked and spotted," which served to distinguish the flocks of the patriarch Jacob from those of Laban his father-in-law, but which did not distinguish goats from sheep, nor sheep from cattle.

The effect of introducing, after this manner, generalizations made altogether irrespective of *rank*, and avowedly without reference to it, into what are inherently and specifically *questions of rank*, admits of a simple illustration.

Let us suppose that it was not with the standing of the Silurian placoids that we had to deal, but with that of the *mammals* of the recent period, including the *quadrumana*, and even the *bimana*, and that we had ventured to describe them, in the words of the Edinburgh Reviewer, as "the very highest types of their class." What would be thought of the reasoner who, in challenging the justice of the estimate, would argue that these creatures, men as well as monkeys, belonged simply to that division of red-blooded animals which includes with the bimana and quadrumana, the frog, the gudgeon, and the *earthworm* ?—a division, he might add, "which, when details of organization are regarded, stretches farther, both downward and upward," than that division of the white-blooded animals to which the crab, the spider, the cuttle-fish, and the dragon-fly belong ; "so that, looking at one extremity, any one is as much entitled to call the red-blooded animals the lowest" division, as any other, looking at another extremity, is to call them the highest division, of animals." What, it might well be asked in reply, has the earthworm, with its red blood, to do in a question respecting the place and standing of the bimana? Or what, in the parallel case, have the *Suctorii*,—the worms of Linnæus,—to do in a question respecting the place and standing of the real placoids? True it is that, according to one principle of classification, now grown somewhat obsolete, men and earthworms are equally red-blooded animals; true it is that, according to another principle of classification, the placoids of Agassiz and the cartilaginous worms of Linnæus are equally *Chondropterygii*. The bimana and the earthworm have their red blood in common; the glutinous hag and the true placoids have as certainly their internal cartilage in common; and if the fact of the red blood of the worm lowers in no degree the rank of the bimana, then, on the same principle, the fact of the internal cartilage of the glutinous hag cannot possibly detract from the standing of the true placoid. In both cases they are creatures that entirely differ,—the earthworms from the bimana, and the cartilaginous *worms* from the placoids; and the classification which tags them together, whether it be that of Aristotle or that of Cuvier, cannot be converted into a sort of minus quantity, of force enough to detract from the value and standing of the bimana in the one case, or of the true placoids in the other. It is in no degree derogatory to the human family that earthworms possess red blood; it is in no degree derogatory to the true placoids that the *Suctorii* possess cartilaginous skeletons.

Let the reader now mark the use which has been made by the author of the "Vestiges," of the name and authority of Linnæus. "Linnæus," he states, "was so impressed by the low character of many of this order (the *Chondropterygii*), that he actually ranked them with worms." Now, what is the fact here? Simply that Linnæus had no such general order as the *Chondropterygii* in his eye at all. Though chiefly remarkable as a naturalist for the artificialness of his classifications, his estimate of the cartilaginous fishes was remarkable,—though carried too far in its extremes, and in some degree founded in error,—for an opposite quality. It was an estimate formed, in the main, on a natural basis. Instead of taking their cartilaginous skeleton into account, he looked chiefly at their standing as animals; and, struck with that extent of front which they present, and with both their superiority on the extreme right, and their inferiority on the extreme left, to the ordinary fishes, he erected them into two separate orders, the one lower and the other higher than the members of the osseous line. And so far was he from regarding the true placoids,—those *Chondropterygii* which to an internal skeleton of cartilage add external plates, points, or spines of bone,—as low in the scale, that he actually raised them above fishes altogether, by erecting them into an order of reptiles,—the order *Amphibia Nantes.* Surely, if the name of Linnæus was to be introduced into this controversy at all, it ought to have been in connection with *this* special fact; seeing that the point to be determined in the question under discussion is simply the place and standing of that very order which the naturalist rated so high,—not the place and standing of the order which he degraded. It so happens that there is one of the *Chondropterygii* which, so far from being a true placoid, does not possess a single osseous plate, point, or spine: it is a worm-like creature, without eyes, without moveable jaws, without vertebral joints, without scales, always enveloped in slime, and greatly abhorred by our Scotch boatmen of the Moray Frith, who hold that it burrows, like the grave-worm, in the decaying bodies of the dead. And this creature, "the glutinous hag," or, according to north-country fishermen, the "ramper-eel," or "poison-ramper," was regarded by Linnæus as belonging, not to the class of fishes, but to the vermes. Now, *this* is the special fact with which, in the development controversy, the author of the "Vestiges" connects the name of the Swedish naturalist! All the fish of the Silurian System belonged to that true placoid order which Linnæus, impressed by its high standing, erected into an order, not of worms, but of reptiles. He elevated A, the true placoid, while he degraded B, the glutinous hag. But it was necessary to the argument of the author of the "Vestiges" that the earliest existing fish should be represented as fish low in the scale; and so he has cited the name and authority of Linnæus in its bearing against the glutinous hag B, as if it had borne against the standing of the true placoid A. The Patagonians are the tallest and bulkiest men in the world, whereas their neighbours the Fuegians are a slim and diminutive race. And if, in some controversy raised regarding the real size of the more gigantic tribe, they were to be described as the "very *tallest* types of their class," any statement in reply, to the effect that some trustworthy voyager had examined certain races of the extreme south of America, and had found that they were both short and thin, would be neither relevant in its facts nor legitimate in its bearing. But if the controversialist who thus strove to strengthen his case by the voyager's authority, was at the same time fully aware that the voyager had seen not only the diminutive Fuegians, but also the gigantic Patagonians, and that he had described these last as very gigantic indeed, the introduction of the statement regarding the smaller race, when he wholly sank the statement regarding the larger, would be not merely very irrelevant in the circumstances, but also very unfair. Such, however, is the style of statement to which the author of the "Vestiges" has (I trust inadvertently) resorted in this controversy.

It is not uninstructive to mark how slowly and gradually the naturalists have been groping their way to a right classification in the ichthyic department of their science, and how it has been that identical perception of resemblance, "having *cartilage* for its general idea, to which the author of the "Vestiges" attaches so much importance, that has served mainly to retard their progress. Not a few of the more distinguished among their number deemed it too important a distinction to be regarded as merely secondary ; and so long as it was retained as a primary characteristic, the fishes failed to range themselves in the natural order ;—dissimilar tribes were brought into close neighbourhood, while tribes nearly allied were widely separated. It failed, as has been shown, to influence Linnæus ; and though he no doubt pressed his peculiar views too far when he degraded the glutinous hag into a worm, and elevated the Sharks and Rays into reptiles, it is certainly worthy of remark that, in the scheme of classification which is now regarded as the *most natural*,—that of Professor Muller, modified by Professor Owen,—the ichthyic worms of the Swede are placed in the first and lowest order of fishes,—the *Dermopteri*,—and the greater part of his ichthyic reptiles, in the eleventh and highest,—the *Plagiostomi*. Cuvier yielded, as has been shown, to the idea of resemblance founded on the *material* of the ichthyic framework, and so ranged his fishes into two parallel lines. Professor Oken, after first enunciating as law that "the characteristic *organ* of fishes is the osseous system," confessed the "great difficulty" which attaches to the question of skeletal "texture or substance," and finally gave up the distinction founded on it as obstinately irreduceable to the purposes of a natural classification.

"The cartilaginous fishes," he says, "appear to belong to each other, and are also usually arranged together ; yet amongst them we find those species, such as the Lampreys, which obviously occupy the lowest grade of all fishes, while the Sharks and Rays remind us of the Reptilia." And so, sinking the consideration of texture altogether, he placed the family of the Lamprey, including the glutinous hag, at the bottom of the scale, and the Sharks and Rays at the top. Agassiz's system, peculiarly his own, has had the rare merit, as I have shown, of furnishing a key to the history of the fish in its several dynasties, which we may in vain seek in any other. His divisions,—if, retaining his strongly-marked placoids and ganoids, as orders stamped in the mint of nature, we throw his perhaps less obviously divisible ctenoids and cycloids into one order,—the corneous or horn-covered,—are scarcely less representative of periods than those great classes of the vertebrata, mammals, birds, reptiles, and fishes, which we find not less regularly ranged in their order of succession in the geologic record than in the "Animal Kingdom" of Cuvier,—a shrewd corroboration, in both cases, I am disposed to think, of the rectitude of the arrangement. What seems to be the special defect of his system is, that having erected his four orders, and then finding a certain number of residuary families that, on his principle of cuticular character, stubbornly refused to fall into any determinate place, he distributed them among the others, with reference chiefly to the totally distinct principle of Cuvier. Thus the *Suctorii*, soft, smooth, slimy-skinned fishes, that do not possess a single placoid character, and are not true placoids, he has yet placed in his placoid order, influenced, apparently, by the "perception of resemblance that has *cartilage* for its central idea ;" and the effect has been a massing into one anomalous and entangled group the fishes of the first period of geologic history, with fishes of which we do not find a trace save in the existing scene of things, and of the highest families of their class with families that occupy the lowest place. But we live in an age in which even the benefactors of the world of mind cannot make false steps with impunity ; and so, while Agassiz's *three* ichthyic orders will continue to be recognised by the palæontologist as the orders of three great geologic periods, the *Suctorii* have already been struck from off his higher fishes by the classification of Muller and Owen, and carried to that lowest point in the scale (indicated by Linnæus and Oken) which their inferior standing renders so obviously the natural one. Some of my readers may perhaps remember how finely Bacon, in his "Wisdom of the Ancients," interprets the old mythologic story of Prometheus. Prometheus, says the philosopher, had conferred inestimable favours on men, by moulding their forms into shape, and bringing them fire from heaven ; and yet they complained of him and his teachings to Jupiter. And the god, instead of censuring their ingratitude, was pleased with the complaint, and rewarded them with gifts. In putting nature to the question, it is eminently wholesome to be doubting, cross-examining, complaining; ever demanding of our masters and benefactors the philosophers, that they should reign over us, not arbitrarily and despotically,

> "Like the old kings, with high exacting looks,
> Sceptred and globed,"

but like our modern constitutional monarchs, who govern by law ; and, further, that an appeal from their decisions on all subjects within the jurisdiction of Nature should for ever lie open to Nature herself. The seeming ingratitude of such a course, if the "complaints" be made in a right spirit and on proper grounds, Jupiter always rewards with gifts.

Let us now see for ourselves, in this spirit, whether there may not be something absolutely derogatory, in the existence of a cartilaginous skeleton, to the creatures possessing it; or whether a deficit of internal bone may not be greatly more than neutralized, as it assuredly must have been in the view of Linnæus, Muller, and Owen, by a larger than ordinary share of a vastly more important substance.

THE PLACOID BRAIN.

EMBRYOTIC CHARACTERISTICS NOT NECESSARILY OF A LOW ORDER.

THAT special substance, according to whose mass and degree of development all the creatures of this world take rank in the scale of creation, is not *bone*, but *brain*. Were animals to be ranged according to the solidity of their bones, the class of birds would be assigned the first place ; the family of the *Felidæ*, including the tiger and lion, the second ; and the other terrestrial carnivora the third. Man and the herbivorous animals, though tolerably low in the scale, would be in advance of at least the reptiles. Most of these, however, would take precedence of the sagacious *Delphinidæ* ; the osseous fishes would come next in order ; the true placoids would follow, succeeded by the *Sturiones ;* and the *Suctorii, i. e.* Cyclostomi or Lampreys, would bring up the rear. There would be evidently no order here : the utter confusion of such an arrangement, like that of the bits of a dissected map flung carelessly out of its box by a child, would of itself demonstrate the inadequacy and erroneousness of the regulating principle. But how very different the appearance presented, when for *solidity of bone* we substitute *development of brain !* Man takes his proper place at the head of creation ; the lower mammalia follow,—each species in due order, according to its modicum of intelligence ; the birds succeed the mammalia ; the reptiles succeed the birds ; the fishes succeed the reptiles ; next in the long procession come the invertebrate animals ; and these, too, take rank, if not according to their development of brain proper, at least according to their development of the *substance* of brain. The occipital nervous ganglion of the scorpion greatly exceeds in size that of the earthworm ; and the occipital nervous ring of the lobster, that of the intestinal ascaris. At length, when we reach the lowest or *acrite* division of the animal kingdom, the substance of brain altogether disappears. It has been calculated by naturalists, that in the vertebrata, the brain in the class of fishes bears an average proportion to the spinal cord of about two to one ; in the class of reptiles, of about two and a half to one ; in the class of birds, of about three to one ; in the class of mammals, of about four to one ; and in the high-placed, sceptre-bearing human family, a proportion of not less than *twenty-three* to one. It is palpably according to development of brain, not development of bone, that we are to determine points of precedence among the animals,—a fact of which no one can be more thoroughly aware than the author of the "Vestiges" himself. Of this let me adduce a striking instance, of which I shall make farther use anon.

"All life," says Oken, "is from the sea ; none from the continent. Man also is a child of the warm and shallow parts of the sea in the neighbourhood of the land." Such also was the hypothesis of Lamarck and Maillet. In following up the view of his masters, the author of the "Vestiges" fixes on the *Delphinidæ* as the sea-inhabiting progenitors of the simial family, and, through the simial family, of man.

For that highest order of the mammalia to which the *Simiadæ* (monkeys) belong, " there remains," he says, " a basis in the *Delphinidæ*, the last and smallest of the cetacean tribes. This affiliation has a special support in the brain of the dolphin family, which is distinctly allowed to be, in proportion to general bulk, the greatest among mammalia next to the ourang-outang and man. We learn from Tiedemann, that each of the cerebral hemispheres is composed, as in man and the monkey tribe, of three lobes,—an anterior, a middle, and a posterior ; and these hemispheres present much more numerous circumvolutions and grooves than those of any other animal. Here it might be rash to found anything upon the ancient accounts of the dolphin,—its familiarity with man, and its helping him in shipwreck and various marine disasters ; although it is difficult to believe these stories to be altogether without some basis in fact. There is no doubt, however, that the dolphin evinces a predilection for human society, and charms the mariner by the gambols which it performs beside his vessel."

Here, then, the author of the " Vestiges" palpably founds on a large development of brain in the dolphin, and on the manifestation of a correspondingly high order of instincts,— and this altogether irrespective of the structure or composition of the creature's internal skeleton. The substance to which he looks as all-important in the case is *brain*, not *bone*. For were he to estimate the standing of the dolphin, not by its brain, but by its skeleton, he would have to assign to it a place, not only *not* in advance of its brethren the *mammalia* of the sea, but even in the rear of the *reptiles* of the sea, the marine tortoises, or turtles,—and scarce more than abreast of the osseous fishes. " Fishes," says Professor Owen, in his " Lectures on the Vertebrate Animals," " have the least proportion of earthy matter in their bones; birds the largest. The mammalia, especially the active, predatory species, have more earth, or harder bones, than reptiles. In each class, however, there are differences in the density of bone among its several members. For example, in the fresh-water fishes, the bones are lighter, and retain more animal matter, than in those which swim in the denser sea. And in the *dolphin*, a warm-blooded marine animal, they differ little in this respect from those of the sea-fish." Such being the fact, it is surely but fair to inquire of the author of the "Vestiges," why he should determine the rank and standing of the *Delphinidæ* according to one set of principles, and the rank and standing of the placoids according to another and entirely different set? If the *Delphinidæ* are to be placed high in the scale, notwithstanding the softness of their skeletons, simply because their brains are large, why are the placoids to be placed low in the scale, notwithstanding the largeness of their brains, simply because their skeletons are soft? It is not too much to demand, that on the principle which he himself recognises as just, he should either degrade the dolphin or elevate the placoid. For it is altogether inadmissible that he should reason on one set of laws when the exigencies of his hypothesis require that creatures with soft skeletons should be raised in the scale, and on another and entirely different set when its necessities demand that they should be depressed.

But do the placoids possess in reality a large development of brain? I have examined the brains of almost all the common fish of our coast, both osseous and cartilaginous, not, I fear, with the skill of a Tiedemann, but all the more intelligently in consequence of what Tiedemann had previously done and written; and so I can speak with some little confidence on the subject, so far at least as my modicum of experience, thus acquired, extends. Of all the common fish of the Scottish seas, the spotted or lesser dog-fish bears, in proportion to its size, the largest brain ; the gray or picked dog-fish ranks next in its degree of development ; the Rays, in their various species, follow after ; and the osseous fishes compose at least the great body of the rear ; while still further behind, there lags a hapless class,—the *Suctorii*, one of which, the glutinous hag, has scarce any brain, and one, the *Amphioxus* or lancelet, wants brain altogether. I have compared the brain of the spotted dog-fish with that of a young alligator, and have found that in scarce any perceptible degree was it inferior, in point of bulk, and very slightly indeed in point of organization, to the brain of the reptile. And the instincts of this placoid family,—one of the truest existing representatives of the placoids of the Silurian System* to which we can appeal,—correspond, we invariably find, with their superior cerebral development. I have seen the common dog-fish, *Spinax Acanthias,* hovering in packs in the Moray Frith, some one or two fathoms away from the side of the herring boat from which, when the fishermen were engaged in hauling their nets, I have watched them, and have admired the caution which, with all their ferocity of disposition, they rarely failed to manifest ;—how they kept aloof from the net, even more warily than the cetacea themselves,—though both dog-fish and cetacea are occasionally entangled ;—and how, when a few herrings were

* The Silurian placoids are *most* adequately represented by the *Cestracion* of the southern hemisphere ; but I know not that of the peculiar character and instincts of this interesting placoid,—the last of its race,—there is anything known. For its form and general appearance see fig. 49, page 153.

shaken loose from the meshes, they at once darted upon them, exhibiting for a moment through the green depths, the pale gleam of their abdomens, as they turned upon their sides to seize the desired morsels,—a motion rendered necessary by the position of the mouth in this family;—and how next, their object accomplished, they fell back into their old position, and waited on as before. And I have been assured by intelligent fishermen, that at the deep-sea white-fishing, in which baited hooks, not nets, are employed, the degree of shrewd caution exercised by these creatures seems more extraordinary still. The hatred which the fisher bears to them arises not more from the actual amount of mischief which they do him, than from the circumstance that in most cases they persist in doing it with complete impunity to themselves. I have seen, said an observant Cromarty fisherman to the writer of these chapters, a pack of dog-fish watching beside our boat, as we were hauling our lines, and severing the hooked fish, as they passed them, at a bite, just a little above the vent, so that they themselves escaped the swallowed hook; and I have frequently lost, in this way, no inconsiderable portion of a fishing. I have observed, however, he continued, that when a fresh pack of hungry dog-fish came up, and joined the pack that had been robbing us so coolly, and at their leisure, a sudden rashness would seize the whole,—the united packs would become a mere heedless mob, and, rushing forward, they would swallow our fish entire, and be caught themselves by the score and the hundred. We may see something very similar to this taking place among even the shrewder mammalia. When pug refuses to take his food, his mistress straightway calls upon the cat, and, quickened by the dread of the coming rival, he gobbles up his rations at once. With the comparatively large development of brain, and the corresponding manifestations of instinct, which the true placoids exhibit, we find other unequivocal marks of a general superiority to their class. In their reproductive organs they rank, not with the common fishes, nor even with the lower reptiles, but with the Chelonians and the Sauria. Among the Rays, as among the higher animals, there are individual attachments formed between male and female: their eggs, unlike the mere spawn of the osseous fishes, or of even the Batrachians, are, like those of the tortoise and the crocodile, comparatively few in number, and of considerable size: their young, too, like the young of birds and of the higher reptiles, pass through no such metamorphosis as those of the toad and frog, or of the amphibia generally. And some of their number,—the common dog-fish for instance,—are ovoviviparous, bringing forth their young, like the common viper and the viviparous lizard, alive and fully formed.

"But such features," says the author of the "Vestiges," referring chiefly to certain provisions connected with the reproductory system in the placoids, "are partly partaken of by families in inferior sub-kingdoms, showing that they cannot truly be regarded as marks of grade in their own class." Nay, single features do here and there occur in the inferior sub-kingdoms, which very nearly resemble single features in the placoid character and organization, which even very nearly resemble single features in the *human* character and organization; but is there any of the inferior sub-kingdoms in which there occurs such a *collocation* of features? or does such a collocation occur in any class of animals,—setting the placoids wholly out of view,—which is not a high class? Nay, farther, does there occur in any of the inferior sub-kingdoms,—existing even as a single feature,—that most prominent, leading characteristic of this series of fishes,—a large brain?

But is not the "cartilaginous structure" of the placoids analogous to the embryotic state of vertebrated animals in general? Do not the other placoid peculiarities to which the author of the "Vestiges" refers,—such as the heterocercal or one-sided tail, the position of the mouth on the under side of the head, and the rudimental state of the maxillaries and intermaxillaries,—bear further analogies with the embryotic state of the higher animals? And is not "embryotic progress the grand key to the theory of development?" Let us examine this matter. "These are the characters," says this ingenious writer, "which, above all, I am chiefly concerned in looking to; for they are features of embryotic progress, and embryotic progress is the grand key to the theory of development." Bold assertion, certainly; but, then, assertion is not argument! The statement is not a reason for the faith that is in the author of the "Vestiges," but simply an avowal of it; it is simply a confession, not a defence, of the Lamarckian creed; and, instead of being admitted as embodying a first principle, it must be put stringently to the question, in order to determine whether it contain a principle at all.

In the first place, let us remark, that the cartilaginous structure of the placoids bears no very striking analogy to the cartilaginous structure of the higher vertebrata in the embryotic state. In the case of the *Delphinidæ*, with their soft skeletons, the analogy is greatly more close. Bone consists of animal matter, chiefly gelatinous, hardened by a diffusion of inorganic earth. In the bones of young and fœtal mammalia, inhabitants of the land, the gelatinous prevails; in the old and middle-aged there is a preponderance of the earth. Now, in the bones of the dolphin there is comparatively little earth. The analogies of its internal skeleton bear, not on the skeletons of its brethren the mature full-grown mammals of the land, but on the skeletons of their immature or fœtal offspring. But in the case of the true placoids that analogy is faint indeed. Their skeletons contain true bone;—the vertebral joints of the Sharks and Rays possess each, as has been shown, an osseous nucleus, which retains, when subjected to the heat of a common fire, the complete form of the joint; and their cranial framework has its surface always covered over with hard osseous points. But though their skeletons possess thus their modicum of bone, unlike those of embryotic birds or mammals, they contain, in what is properly their cartilage, no gelatine. The analogy signally fails in the very point in which it has been deemed specially to exist. The cartilage of the *Chondropterygii* is a substance so essentially different from that of young or embryotic birds and mammals, and so unique in the animal kingdom, that the heated water in which the one readily dissolves has no effect whatever upon the other. It is, however, a curious circumstance, exemplified in some of the Shark family,* though it merely serves, in its exceptive character, to establish the general fact, that while the rays of the double fins, which answer to the phalanges, are all formed of this *indissoluble* cartilage, those rays which constitute their outer framework, with the rays which constitute the framework of all the single fins, are composed of a *mucoidal* cartilage, which boils into glue. At certain definite lines a change occurs in the texture of the skeleton; and it is certainly suggestive of thought, that the difference of substance which the change involves distinguishes that

* Such as the dog-fishes, picked and spotted.

part of the skeleton which is homologically representative of the skeletons of the higher vertebrata, from that part of it which is peculiar to the creature as a fish, viz. the dorsal and caudal rays, and the extremities of the double fins. These emphatically ichthyic portions of the animal may be dissipated by boiling, whereas what Linnæus would perhaps term its *reptilian* portion abides the heat without reduction.

But is not the one-sided tail, so characteristic of the sharks, and of almost all the ancient ganoids, also a characteristic of the young salmon just burst from the egg? Yes, assuredly; and, so far as research on the subject has yet extended, of not only the salmon, but of *all* the other osseous fishes in their fœtal state. The salmon, on its escape from the egg, is a little monster of about three quarters of an inch in length, with a huge heart-shaped bag, as bulky as all the rest of its body, depending from its abdomen. In this bag provident nature has packed up for it, in lieu of a nurse, food for five weeks; and, moving about everywhere in its shallow pool, with its provision knapsack slung fast to it, it reminds one disposed to be fanciful, save that its burden is on the wrong side, of Scottish soldiers of the olden time summoned to attend their king in war,—

> "Each on his *back*, a slender store,
> His forty days' provision bore,
> As ancient statutes tell."

Around that terminal part of the creature's body traversed by the caudal portion of the vertebral column, which commences in the salmon immediately behind the ventrals, there runs at this period, and for the ensuing five weeks in which it does not feed, a membranous fringe or fin, which exactly resembles that of the tadpole, and which, existing simply as an expansion of the skin, exhibits no mark of rays. In the place of the true caudal fin, however, we may detect, with the assistance of a lens, an internal framework with two well-marked lobes, and ascertain, further, that this tail is set on awry,—the effect of a slight upward bend in the creature's body. And when viewed in a strong light as a transparency, we perceive that the spinal cord takes the same upward bend, and, as in the sturgeon, passes in an exceedingly attenuated form into the upper lobe. What may be regarded as the *design* of the arrangement is probably to be found in the peculiar form given to the little creature by the protuberant bag in front. A wise instinct teaches it, from the moment of its exclusion from the egg, to avoid its enemies. In the instant the human shadow falls upon its pool, we see it darting into some recess at the sides or bottom, with singular alacrity; and in order to enable it to do so, and to steer itself aright,—as, like an ill-trimmed vessel, deep in the water a-head, the balance of its body is imperfect,—there is, if I may so express myself, a heterocercal peculiarity of helm required. It has got an irregularly-developed tail to balance an irregularly-developed body, as skiffs *lean* on the one beam and *full* on the other require, in rowing, a cast of the rudder to keep them straight in their course.

Sinking altogether, however, the final cause of the peculiarity, and regarding it simply as a *fœtal* one, that indicates a certain stage of imperfection in the creature in which it occurs, on what principle, I ask, are we to infer that what is a sign of immaturity in the young of one set of animals, is a mark of inferior organization in the adult forms of another set? The want of eyes in any of the animal families, or the want of organs of progression, or a fixed and sedentary condition, like that of the oyster, are all marks of great inferiority. And yet, if we admit the principle, that what are evidences of immaturity in the young members of one family are signs of inferior organization in the fully-grown members of another, it could easily be shown that eyes and legs are defects, and that the unmoving oyster stands higher in the scale than the ever-restless fish or bird. The immature *Tubularia* possess locomotive powers, whereas in their fully developed state they remain fixed to one spot in their convoluted tubes. The immature *Lepas* is furnished with members well adapted for swimming, and with which it swims freely; as it rises towards maturity, these become blighted and weak; and, when fully grown,—fixed by its fleshy pedicle to the rock or floating log to which it attached itself in its transition state,—it is no longer able to swim. The immature *Balanus* is furnished with two eyes: in its state of maturity these are extinguished, and it passes its period of full development in darkness. Further, it is not generally held that in the human family a white skin is a decided mark of degradation, but rather the reverse; and yet nothing can be more certain than that the Negro fœtus has a white skin. Since eyes, and organs of progression, and a power of moving freely, and a white skin, are mere embryotic peculiarities in the *Balanus*, the *Lepas*, the *Tubularia*, and the Negro, and yet are in themselves, when found in the mature animal, evidences of a high, not of a low standing, on what principle, I ask, are we to infer that the peculiarity of a heterocercal tail, embryotic in the salmon, is, when found in the mature placoid, an evidence, not of a high standing, but of a low? Every true analogy in the case favours an exactly opposite view. In the heterocercal or one-sided tail, the vertebral joints gradually diminish, as in the tails of the *Sauria* and *Ophidia*, till they terminate in a point; whereas the homocercal tail common to the osseous fishes exhibits no true analogy with the tails of the higher orders. Its abruptly terminating vertebral column, immensely developed posterior processes, and broadly expanded osseous rays, seem to be simply a few of the many marks of decline and degradation which fishes, the oldest of the vertebrata, exhibit in this late age of the world, and which, in at least the earlier geologic periods, when they were greatly younger as a class, they did not betray.

Fig. 48.

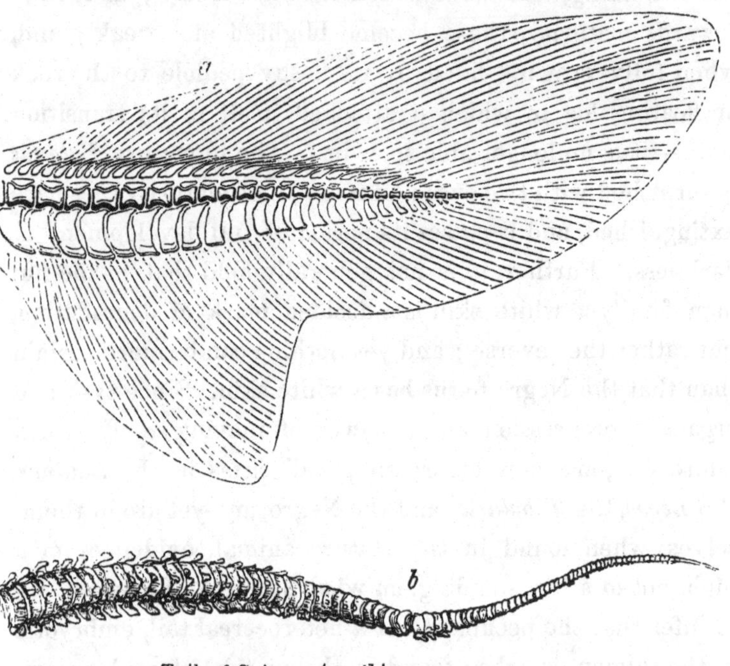

a. *Tail of Spinax Acanthias.*
b. *Tail of Ichthyosaurus Tenuirostris* (Buckland).

In illustration of this view, I would fain recommend to the reader a simple experiment. Let him procure the tail of a common dog-fish (fig. 48 *a*), and, cutting it across about half an inch above where the caudal fin begins, let him boil it smartly for about half an hour. He will first see it swell, and then burst, all around those thinner parts of the fin that are traversed by the caudal rays,—wholly mucoidal, as shown by this test, in their texture, and which yield to the boiling water, as if formed of isinglass. They finally dissolve, and drop away, with the surrounding cuticular integument ; and then there only remains, as the insoluble framework of the whole, the bodies of the vertebræ, with their neural and hœmal processes. The tail has now lost much of its ichthyic character, and has acquired, instead, a considerable degree of resemblance to the reptilian tail, as exemplified in the saurians. I have introduced into the wood-cut, for the purpose of comparison, the tail of the ichthyosaurus (*b*). It consists, like the other, of a series of gradually diminishing vertebræ, and must have also supported, says Professor Owen, a propelling fin, placed vertically, as in the shark, which, however, from its perishable nature, has in every instance disappeared in the earth, as that of the dog-fish disappears in the boiling water. It will be seen that its processes are comparatively smaller than those of the fish, and that the bodies of its vertebræ are shorter and bulkier ; but there is at least a general correspondence of the parts; and were the tail of the crocodile, of which the vertebral bodies are slender and the processes large,to be substituted for that of the enaliosaur here, the correspondence would be more marked still. After thus *developing* the tail of the reptile out of that of the fish,—as the cauldron-bearing Irish magician of the tale developed young ladies out of old women,—simply by *boiling*, let the reader proceed to a second stage of the experiment, and see whether he may not be able still farther to develope the reptilian tail so obtained, into that of the mammal, by *burning*. Let him spread it out on a piece of iron hoop, and thrust it into the fire ; and then, after exposure for some time to a red heat has consumed and dissipated its merely cartilaginous portions, such as the neural and hœmal processes, with the little pieces which form the sides of the neural arch, and left only the whitened bodies of the vertebræ, let him say whether the bony portion which remains does not present a more exact resemblance to the mammiferous tail,—that of the dog, for example,—than anything else he ever saw. The Lamarckians may well deem it an unlucky circumstance, that one special portion of their theory should demand the depreciation of the heterocercal tail, seeing that it might be represented with excellent effect in another, as not merely a connecting link in the upward march of progression between the tail of the true fish and that of the true reptile, but as actually containing in itself,—as the caterpillar contains the future pupa and butterfly,—the elements of the reptilian and mammiferous tail. If there be any virtue in analogy, the heterocercal tail is, I repeat, of a decidedly higher type than the homocercal one. It furnishes the first example in the vertebrata of those coccygeal vertebræ diminishing to a point, which characterizes not only all the higher reptiles, but also all the higher mammals, and which we find represented by the *Os coccygis* in man himself. But to this special point I shall again refer.

With regard to that rudimentary state of the *occipital* framework of the placoids to which the author of the "Vestiges" refers, it may be but necessary to say that, notwithstanding the simplicity of their box-like skulls, they bear in their character, as cases for the protection of the brain, at least as close an analogy to the skulls of the higher animals, as those of the osseous fishes, which consist usually of the extraordinary number of from sixty to eighty bones,—a mark,—the author of the "Vestiges" himself being judge in the case,—rather of inferiority than the reverse. "Elevation is marked in the scale," we find him saying, "by an animal exchanging a multiplicity of parts serving one end, for a smaller number." The skull of a cod consists of about thrice as many separate bones as that of a man. But I do not well see that in this case the fact either of *simplicity* in excess or of *multiplicity* in excess can be insisted upon in either direction, as a proper basis for argument. Nearly the same remark applies to the maxillaries as to the skull. The under jaw in man consists of a single bone; that of the thornback,—if we do not include the two suspending *ribs*, which belong equally to the upper jaw,—of two bones (the number in all the mammiferous quadrupeds); that of the cod, of four bones, and, if we include the suspending *ribs*, of twelve. On what principle are we to hold, with *one* as the representative number of the highest type of jaw, that *two* indicates a lower standing than *four*, or *four* than *twelve?* In reference to the further statement, that in many of the ancient fishes "traces can be observed of the muscles having been attached to the external plates, strikingly indicating their low grade as vertebrate animals," it may be answer enough to state, that the peculiarity in question was not a characteristic of the *most* ancient fishes,—the placoids of the Silurian system,—but of some ganoids of the succeeding systems. The reader may remember, as a case in point, the example furnished by the nail-like bone of *Asterolepis*, figured in page 87, in which there exists depressions resembling that of the round ligament in the head of the quadrupedal thigh-bone. And as for the remark that the opening of the mouth of the placoid "on the under side of the head," is indicative of a low embryotic condition, it might be almost sufficient to remark, in turn, that the lowest family of fishes,—that to which the supposed worms of Linnæus belong,—have the mouth not under, but at the anterior termination of the head,—in itself an evidence that the position of the mouth at the extremity of the muzzle, common to the greater number of the osseous fishes, can be no very high character, seeing that the humblest of the *Suctorii* possess it ; and that many osseous fishes, whose mouths open, not on the under, but the upper side of the snout, as in the distorted and asymetrical genus *Platessa*, are not only in no degree superior to their bony neighbours, and far inferior to the placoid ones, but bear, in direct consequence of the arrangement, an expression of unmistakeable stupidity. The objection, however, admits of a greatly more conclusive reply.

"This fish, to speak in the technical language of Agassiz," says the Edinburgh Reviewer, in reference to the ancient ichthyolite of the Wenlock Shale, "undoubtedly belongs to the Cestraciont family of the placoid order,—proving to demonstration that the oldest known fossil fish [1845] belongs to the highest type of that division of the vertebrata." I may add, that the character and family of this ancient specimen was determined by our highest British authority in fossil ichthyology, Sir Philip Egerton. And it is in depreciation of Professor Sedgwick's statement regarding its high standing that the author of the "Vestiges" refers to the supposed inferiority indicated by a mouth opening, not at the extremity of the muzzle, but under the head. Let us, then, fully grant, for the argument's sake, that the occurrence of the mouth in the muzzle *is* a sign of superiority, and its occurrence under the head a mark of great inferiority, and then ascertain how the fact stands with regard to the *Cestracion*. "The Cestracion sub-genus," says Mr James Wilson, in his admirable treatise on fishes, which forms the article

ICHTHYOLOGY in the "Encylopædia Britannica," "has the temporal aperture, the anal fin, and rounded teeth, of *Squalus Mustelus ; but the mouth is* TERMINAL, *or* AT THE EXTREMITY OF THE POINTED MUZZLE." The accompanying figure (49),

Fig. 49.

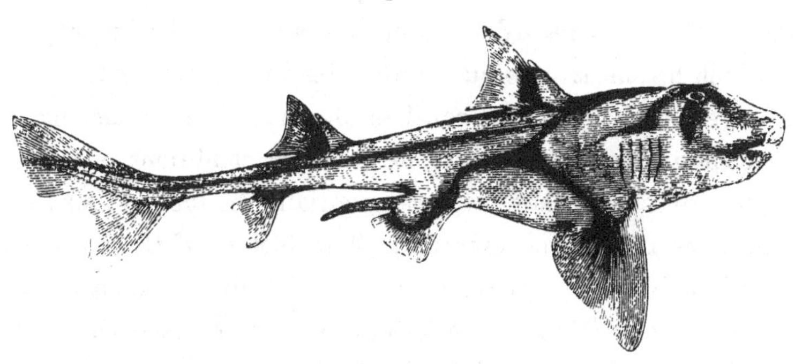

PORT-JACKSON SHARK (*Cestracion Phillippi.*)

taken from a specimen of *Cestracion* in the collection of Professor John Fleming, may be regarded as of some little interest, both from its direct bearing on the point in question, and from the circumstance that it represents, not inadequately for its size, the sole surviving species (*Cestracion Phillippi*) of the oldest vertebrate family of creation. With this family, so far as is yet known, ichthyic existence first began. It does not appear that on the globe which we inhabit there was ever an ocean tenanted by living creatures at all that had not its *Cestracion*,—a statement which could not be made regarding any other vertebrate family. In Agassiz's "Tabular View of the Genealogy of Fishes," the Cestracionts, and they only, sweep across the entire geologic scale. And, as shown in the figure, the mouth in this ancient family, instead of opening, as in the ordinary sharks, under the middle of the head, to expose them to the suspicion of being creatures of low and embryotic character, opened in a broad honest-looking muzzle, very much resembling that of the hog. The mouths of the most ancient placoids of which we know anything *did not*, I reiterate, *open under their heads.*

But why introduce the element of embryotic progress into this question at all? It is not a question of embryotic progress. The very legerdemain of the sophist,—the juggling by which he substitutes his white balls for black, or converts his pigeons into crows,—consists in the art of attaching the conclusions founded on the facts or conditions of one subject, to some other subject essentially distinct in its nature. Gestation is not creation. The history of the young of animals in their embryotic state is simply the history of the fœtal young; just as the history of insect transformation, in which it has been held by good men but weak reasoners that there exists direct evidence of the doctrine of the Resurrection, is the history of insect transformation, and of nothing else. True, the human mind is so constituted that it converts all nature into a storehouse of comparisons and analogies; and this fact of the metamorphosis of the creeping caterpillar, after first passing through an intermediate period of apparent death as an inert aurelia, into a winged imago, seems to have seized on the human fancy at a very early age, as wonderfully illustrative of life, death, and the future state. The Egyptians wrapped up the bodies of their dead in the chrysalis form, so that a mummy, in their apprehension, was simply a human pupa, waiting the period of its enlargement; and the Greeks had but one word in their language for butterfly and the soul. But not the less true is it, notwithstanding, that the facts of insect transformation furnish no legitimate key to the totally distinct facts of a resurrection of the body, and of a life after death. And on what principle, then, are we to trace the origin of past dynasties in the changes of the fœtus, if not the rise of the future dynasty in the transformations of the caterpillar? "These [embryotic] characters [that of the heterocercal tail, and of the mouth of the ordinary shark type] are essential and important," remarks the author of the "Vestiges," "whatever the Edinburgh Reviewer may say to the contrary;—they are the characters which, above all, I am chiefly concerned in looking to, for they are the features of embryotic progress, and embryotic progress is the grand key to the theory of development." Yes; the grand key to the theory of *fœtal* development; for embryotic progress *is* fœtal development. But on what is the assertion based that they form a key to the history of creation? Aurelia are not human bodies laid out for the sepulchre, nor are butterflies human souls; as certainly gestation is not creation, nor a life of months in the uterus a succession of races for millions of ages outside of it. On what grounds, then, is the assertion made? Does it embody the result of a discovery, or announce the message of a revelation? Did the author of the "Vestiges" find it out for himself, or did an angel from heaven tell it him? If it be a discovery, show us, we ask, the steps through which you have been conducted to it; if a revelation, produce, for our satisfaction, the evidence on which it rests. For we are not to accept as data in a question of science, idle comparisons or vague analogies, whether produced through the intentional juggling of the sophist, or involuntarily conjured up in the dreamy delirium of an excited fancy.

It is one of the difficulties incident to the task of replying to any dogmatic statement of error, that every mere annunciation of a false fact or false principle must be met by elaborate counter-statement or carefully constructed argument, and that prolixity is thus unavoidably entailed on the controversialist who labours to set right what his antagonist has set wrong. The promulgator of error may be lively and entertaining, whereas his pains-taking confutator runs no small risk of being tedious and dull. May I, however, solicit the forbearance of the reader, if, after already spending much time in skirmishing on ground taken up by the enemy,—one of the disadvantages incident to the mere defendant in a controversy of this nature,—I spend a little more in indicating what I deem the proper ground on which the standing of the earlier vertebrata should be decided. To the test of *brain* I have already referred as all-important in the question : I would now refer to the test of what may be termed *homological symmetry of organization*.

THE PROGRESS OF DEGRADATION.

ITS HISTORY.

THOUGH all animals be fitted by nature for the life which their instincts teach them to pursue, naturalists have learned to recognise among them certain aberrant and mutilated forms, in which the type of the special class to which they belong seems distorted and degraded. They exist as the monster *families* of creation, just as among families there appear from time to time monster *individuals*,—men, for instance, without feet, or hands, or eyes, or with their feet, hands, or eyes grievously misplaced,—sheep with their fore legs growing out of their necks, or ducklings with their wings attached to their haunches. Among these degraded races, that of the footless serpent, which "goeth upon its belly," has been long noted by the theologian as a race typical, in its condition and nature, of an order of hopelessly degraded beings, borne down to the dust by a clinging curse; and, curiously enough, when the first comparative anatomists in the world give *their* readiest and most prominent instance of degradation among the denizens of the natural world, it is this very order of footless reptiles that they select. So far as the geologist yet knows, the Ophidians did not appear during the Secondary ages, when the monarchs of creation belonged to the reptilian division, but were ushered upon the scene in the times of the Tertiary deposits, when the mammalian dynasty had supplanted that of the Iguanodon and Megalosaurus. Their ill omened birth took place when the influence of their house was on the wane, as if to set such a stamp of utter hopelessness on its fallen condition, as that set by the birth of a worthless or idiot heir on the fortunes of a sinking family. The degradation of the Ophidians consists in the absence of limbs,—an absence total in by much the greater number of their families, and represented in others, as in the boas and pythons, by mere abortive hinder limbs concealed in the skin; but they are thus not only *monsters through defect of parts*, if I may so express myself, but also *monsters through redundancy*, as a vegetative repetition of vertebra and ribs to the number of three or four hundred forms the special contrivance by which the want of these is compensated. I am also disposed to regard the poison-bag of the venomous snakes as a mark of degradation;—it seems, judging from analogy, to be a protective provision of a low character, exemplified chiefly in the invertebrate families,—ants, centipedes, and mosquitos,—spiders, wasps, and scorpions. The higher carnivora are, we find, furnished with unpoisoned weapons, which, like those of civilized man, are sufficiently effective, simply from the excellence of their construction, and the power with which they are wielded, for every purpose of assault or defence. It is only the squalid savages and degraded boschmen of creation that have their feeble teeth and tiny stings steeped in venom, and so made formidable. *Monstrosity through displacement of parts* constitutes yet another form of degradation; and this form, united, in some instances, to the other two, we find curiously exemplified in the geological history of the fish,—a history which, with all its blanks and missing portions, is yet better known than that of any other division of the vertebrata. And it is, I am convinced, from a survey of the progress of degradation in the great ichthyic division,—a progress recorded as "with a pen of iron in the rock for ever," and not from superficial views founded on the cartilaginous or non-cartilaginous texture of the ichthyic skeleton, that the standing of the kingly fishes of the earlier periods is to be adequately determined. Any other mode of survey, save the parallel mode which takes development of brain into account, evolves, we find, nothing like principle, and lands the inquirer in inextricable difficulties and inconsistencies.

In all the higher non-degraded vertebrata we find a certain uniform type of skeleton, consisting of the head, the vertebral column, and four limbs; and these last, in the various symmetrical forms, whether exemplified in the higher fish, the higher reptiles, the higher birds, the higher mammals, or in man himself, occur always in a certain determinate order. In all the mammals, the scapular bases of the fore limbs begin opposite the eighth vertebra from the skull backwards, the seven which go before being cervical or neck vertebræ; in the birds,—a division of the vertebrata that, from their peculiar organization, require longer and more flexible necks than the mammals,—the scapulars begin at distances from the occiput varying, according to the species, from opposite the thirteenth to opposite the twenty-fourth vertebra; and in the reptiles, —a division which, according to Cuvier, "presents a greater diversity of forms, characters, and modes of gait, than any of the other two,"—they occur at almost all points, from opposite the second vertebra, as in the frog, to opposite the thirty-third or thirty-fourth vertebra, as in some species of plesiosaurus.

But in all,—whether mammals, birds, or undegraded reptiles, —they are so placed, that the creatures possess *necks*, of greater or less length, as an essential portion of their general type. The hinder limbs have also in all these three divisions of the animal kingdom their typical place. They occur opposite, or very nearly opposite, the posterior termination of the abdominal cavity, and mark the line of separation between the vertebræ of the trunk (dorsal, lumbar, and sacral), and the third and last, or *caudal* division of the column,—a division represented in man by but four vertebræ, and in the crocodile by about thirty-five, but which is found to exist, as I have already said, in all the more perfect forms. The limbs, then, in all the symmetrical animals of the first three classes of the vertebrata, mark the three great divisions of the vertebral column,—the division of the *neck*, the division of the *trunk*, and the division of the *tail*. Let us now inquire how the case stands with the fourth and lowest class, —that of the fishes.

In those existing placoids that represent the fishes of the earliest vertebrate period, the places of the double fins,—pectorals and ventrals,—which form in the ichthyic class the true homologues of the limbs, correspond to the places which these occupy in the symmetrical mammals, birds, and reptiles. The scapular bases of the fore or pectoral fins ordinarily begin opposite the twelfth or fourteenth vertebra;* but they range, as in man and the mammals, in a forward direction, so that the fins themselves are opposite the eighth or tenth. The pelvic bases of the ventral fins are placed nearly opposite the base of the abdomen, so that, as in all the symmetrical animals, the

* The twelfth in *Spirax Acanthias*, and the fourteenth in *Scyllium Stellare*.

vent opens between, or nearly between, those hinder limbs which the bases support. In the Rays, which, so far as is yet known, did not appear in creation until the Secondary ages had begun, the bases of the fore limbs, *i.e.* pectoral fins, are attached to the lower part of a huge cervical vertebra, nearly equal in length to *all* the trunk vertebræ united; and in the Chimeridæ, which also first appear in the Secondary division, they are attached, as in the osseous fishes, to the hinder part of the head. But in the representatives of all those Silurian placoids yet known, of which the family can be determined, or anything with safety predicated, the cervical division is found to occur as a series of vertebræ: they present in this, as in the hinder portion of their bodies, the homological symmetry of organization typical of that vertebral sub-kingdom to which they belong.

In the second great period of ichthyic existence,—that of the Old Red Sandstone,—we find the first example, in the class of fishes, of "monstrosity through *displacement* of parts," and apparently also,—in at least two genera, though the evidence on this head be not yet quite complete,—of "monstrosity through *defect* of parts." In all the ganoids of the period, with (so far as we can determine the point) only two exceptions, the scapular bases of the fore limbs are brought forward from their typical place opposite the base of the cervical vertebræ, and stuck on to the occipital plate. There occurs, in consequence, in one great order of the ichthyic class, such a departure from the symmetrical type as would take place in a monster example of the human family in whom the neck had been annihilated, and the arms stuck on to the back of the head. And in the genera *Coccosteus* and *Pterichthys* we find the first example of degradation through *defect*. In the *Pterichthys* the *hinder* limbs seem wanting; and in the *Coccosteus* we find no trace of the *fore* limbs. The one resembles a monster of the human family born without hands, and the other a monster born without feet. Ages and centuries pass, and long unreckoned periods come to a close; and then, after the termination of the Palæozoic period, we see that change taking place in the form of the ichthyic tail, to which I have already referred (and to which I must refer at least once more), as singularly illustrative of the progress of degradation. Yet other ages and centuries pass away, during which the reptile class attains to its fullest development, in point of size, organization, and number; and then, after the times of the Cretaceous deposits have begun, we find yet another remarkable monstrosity of displacement introduced among all the fishes of one very numerous order, and among no inconsiderable proportion of the fishes of another. In the newly-introduced ctenoids (*Acanthopterygii*), and in those families of the cycloids which Cuvier erected into the order *Malacopterygii sub-brachiati*, the hinder limbs are brought forward, and stuck on to the base of the previously misplaced fore limbs. All the four limbs, by a strange monstrosity of displacement, are crowded into the place of the extinguished neck. And such, at the present day, is the prevalent type among fishes. Monstrosity through *defect* is also found to increase; so that the snake-like *apoda*, or feet-wanting fishes, form a numerous order, some of whose genera are devoid, as in the common eels and the congers, of only the hinder limbs; while in others, as in the genera Muræna and Synbranchus, both hinder and fore limbs are wanting. In the class of fishes, as fishes now exist, we find many more evidences of the monstrosity which results from both the misplacement and defect of parts, than in the other three classes of the vertebrata united; and knowing their geological history better than that of any of the others, we know, in consequence, that the monstrosities did not appear *early*, but *late*, and that the progress of the race, as a whole, though it still retains not a few of the higher forms, has been a progress, not of development from the low to the high, but of degradation from the high to the low.

The reader may mark for himself, in the flounder, plaice, halibut, or turbot,—fishes of a family of which there appears no trace in the earlier periods,—an extreme example of the degradation of distortion superadded to that of displacement. At a first glance the *limbs* seem but to exhibit merely the amount of natural misarrangement and misorder common to the *Acanthopterygii* and *Sub-brachiati* ;—the base of the pectorals are stuck on to the head, and the base of the ventrals attached to that of the pectorals. From the circumstance, however, that the creature is twisted half round and laid on its side, we find that at least one of the pairs of double fins,—the pectorals,—perform the part of single fins, —one projecting from the animal's superior, the other from its inferior side, in the way the anal and dorsal fins project from the upper and under surfaces of other fishes ; while its real dorsal and anal fins, both developed very largely, and,— in order to preserve its balance,—in about an equal degree, and wonderfully correspondent in form, perform, from their lateral position, the functions of single fins. Indeed, at a first glance they seem the analogues of the largely-developed pectorals of a very different family of flat fishes,—the Rays. It would appear as if single and double fins, by some such mutual agreement as that which, according to the old ballad, took place between the churl of Auchtermuchty and his wife, had agreed to exchange callings, and perform each the work of the other. The tail, too, possesses, in consequence of the twist, not the vertical position of other fish-tails, but is spread out horizontally, like the tails of the cetacea. It is, however, in the head of the flounder and its cogeners that we find the more extraordinary distortions exemplified. In order to accommodate it to the general twist, which rendered lateral what in other fishes is dorsal and abdominal, and dorsal and abdominal what in other fishes is lateral, one-half its features had to be twisted to the one side, and the other half to the other. The face and cranium have undergone such a change as that which the human face and cranium would undergo, were the eyes to be drawn towards the left ear, and the mouth towards the right. The skull, in consequence, exhibits, in its fixed bones, a strange Cyclopean character, unique among the families of creation: it has its one well-marked eye-orbit opening, like that of Polyphemus, direct in the middle of the fore part of its head; while the other, external to the cranium altogether, we find placed among the free bones, directly over the maxillaries. And the wry mouth,—twisted in the opposite direction, as if to keep up such a balance of deformity as that which the breast-hump of a hunchback forms to the hump behind,—is in keeping with the squint eyes. The jaws are strangely asymmetrical. In symmetrical fishes the two bones that compose the anterior half of the lower jaw are as perfectly correspondent in form and size as the left hand or left foot is correspondent, in the human subject, to the *right* hand or *right* foot; but not such their character in the flounder. The one is a broad, short, nearly straight bone; the other is larger, narrower, and bent like a bow; and while the one contains only from four to six teeth, the other contains from thirty to thirty-five. Scarcely in the entire ichthyic kingdom are there any two jaws that less resemble one another than the two halves of the jaw of the flounder, turbot, halibut, or plaice. The intermaxillary bones are equally ill matched: the one is fully twice the size of the other, and contains about thrice as many teeth. That bilateral symmetry of the skeleton which is so *invariable* a characteristic of the vertebrata, that ordinary observers, who have eyes for only the rare and the uncommon, fail to remark it, but which a Newton could regard as so wonderful, and so thoroughly in harmony with the uniformity of the planetary system, has scarce any place in the asymmetrical head of the flounder. There exists in some of our north country fishing villages an ancient apologue, which, though not remarkable for point or meaning, at least serves to show that this peculiar example of distortion the rude fishermen of a former age were observant enough to detect. Once on a time the fishes met, it is said, to elect a king; and their choice fell on the herring. "The herring king!" contemptuously exclaimed the flounder, a fish of consummate vanity, and greatly piqued on this occasion that its own presumed claims should have been overlooked; "where, then, am I?" And straightway, in punishment of its conceit and rebellion, "its eyes turned to the back of its head." Here is there a story palpably founded on the degradation of misplacement and distortion, which originated ages ere the naturalist had recognised either the term or the principle.

It would be an easy matter for an ingenious theorist, not much disposed to distinguish between the minor and the master laws of organized being, to get up quite as unexceptionable a theory of degradation as of development. The one-eyed, one-legged Chelsea pensioner, who had a child, unborn at the time, laid to his charge, agreed to recognise his relationship to the little creature, if, on its coming into the world, it was found to have a green patch over its eye, and a wooden leg. And, in order to construct a hypothesis of progressive degradation, the theorist has but to take for granted the transmission to other generations of defects and compensating redundancies at once as extreme and accidental as the loss of eyes or limbs, and the acquisition of timber legs or green patches. The snake, for instance, he might regard as a saurian, that, having accidentally lost its limbs, exerted itself to such account throughout a series of generations, in making up for their absence, as to spin out for itself, by dint of writhing and wriggling, rather more than a hundred additional vertebræ, and to alter, for purposes of greater flexibility, the structure of all the rest. And as fishes, when nearly stunned by a blow, swim for a few seconds on their side, he might regard the flounders as a race of half-stunned fishes, previously degraded by the misplacement of their limbs, that, instead of recovering themselves from the blow given to some remote parent of the family, had expended all their energies in twisting their mouths round to what chanced to be the under side on which they were laid, and their eyes to what chanced to be the upper, and that made their pectorals serve for anal and dorsal fins, and their anal and dorsal fins serve for pectorals. But while we must recognise in nature certain laws of disturbance, if I may so speak, through which, within certain limits, traits which are the result of habit or circumstance in the parents are communicated to their offspring, we would err as egregiously, did we take only these into account, without noting that infinitely stronger antagonist law of reproduction and restoration which, by ever gravitating towards the original type, preserves the integrity of races, as the astronomer would, who, in constructing his orrery, recognised only that law of propulsion through which the planets speed through the heavens, without taking into account that antagonist law of gravitation which, by maintaining them in their orbits, ensures the regularity of their movements. The law of restoration would recover and right the stunned fish laid on its side; the law of reproduction would give limbs to the offspring of the mutilated saurian. We have evidence, in the extremeness of the degradation in these cases, that it cannot be a degradation hereditarily derived from accident. Nature is, we find, active, not in perpetuating the accidental wooden legs and green patches of ancestors in their descendants, but in restoring to the offspring the true limbs and eyes which the parents have lost. It is, however, not with a theory of hereditary degradation, but a hypothesis of gradual development, that I have at present to deal; and what I have to establish as proper to the present stage of my argument is, that this principle of degradation really exists, and that the history of its progress in creation bears directly against the assumption that the earlier vertebrata were of a lower type than the vertebrata of the same ichthyic class which exist now.*

---

* It will scarce be urged against the degradation theory, that those races which, tried by the tests of defect or misplacement of parts, we deem degraded, are not less fitted for carrying on what in their own little spheres is the proper business of life, than the non-degraded orders and families. The objection is, however, a possible one, and one which a single remark may serve to obviate. It is certainly true that the degraded families *are* thoroughly fitted for the performance of all the work given them to do. They greatly increase when placed in favourable circumstances, and, when vigorous and thriving, enjoy existence. But then the same may be said of all animals, without reference to their place in the scale;—the mollusc is as thoroughly adapted to its circumstances, and as fitted to accomplish the end proper to its being, as the mammiferous quadruped, and the mammiferous quadruped as man himself; but the fact of perfect adaptation in no degree invalidates the other not less certain fact of difference of rank, nor proves that the mollusc is equal to the quadruped, or the

The progress of the ichthyic tail, as recorded in geologic history, corresponds with that of the ichthyic limbs. And as in the existing state of things we find fishes that *nearly* represent, in this respect, all the great geologic periods,—I say *nearly*, not *fully*, for I am acquainted with no fish adequately representative of the period of the Old Red Sandstone,—it may be well to cast a glance over the *contemporary* series, as illustrative of the *consecutive* one. In those placoids of the shark family that to a large brain unite homological symmetry of organization, and represent the fishes of the first period, we find, as I have already shown, that the vertebræ gradually diminish in the caudal division of the column, until they terminate in a point,—a circumstance in which they resemble not merely the betailed reptiles, but also all the higher mammiferous quadrupeds, and even man himself. And it is this peculiarity, stamped upon the less destructible portions of the framework of the tail,—vertebræ and processes,—rather than the one-sided or heterocercal form of the surrounding fin, composed of but a mucoidal substance, that constitutes its grand characteristic; seeing that in some placoid genera, such as *Scyllium Stellare*, the terminal portion of the fin is scarce less largely developed above than below, and that in others, as in most of the Ray family, the under lobe of the fin is wholly wanting. In the sturgeon,—one of the few ganoids of the present time,—we become sensible of a peculiar modification in this heterocer-

---

quadruped to man. And, of course, the remark equally bears on the *reduced* as on the *unelevated*,—on lowness of place when a result of degradation in races pertaining to a higher division of animals, as on lowness of place when a result of the humble standing of the division to which the races belong.

cal type of tail : the lower lobe is, we find, composed, as in *Spinax* and *Scyllium*, of rays exclusively ; while through the centre of the upper lobe there runs an acutely angular patch of lozenge-shaped plates, like that which runs through the centre of the double fins of *Dipterus* and the Celacanths. But while in the sharks the gradually diminishing vertebræ stand out in bold relief, and form the thickest portion of the tail, that which represents them in the sturgeon (the angular patch) is slim and thin,—slimmer in the middle than even at the sides ;—in part a consequence, no doubt, of the want, in this fish, of solid vertebræ, but a consequence also of the extreme attenuation of the nervous cord, in its prolongation into the lobe of the fin. Further, the rays of the tail,—its peculiarly ichthyic portion, which are purely mucoidal in *Spinax*, *Scyllium*, and *Cestracion*,—have become osseous in the sturgeon. The *fish* has *set* and become *fixed*, as cement sets in a building, or colours are fixed by a mordant. And it is worthy of special remark that, correspondent with the peculiarly *ichthyic* development of tail in this fish, we find the prevailing ichthyic displacement of the fore limbs. Again, in the *Lepidosteus*, another of the true ganoids which still exist, the internal angle of the upper lobe of the tail wholly disappears, and with the internal angle the prolongation of the nervous cord. Still, however, it is what the tail of the sturgeon would become were the angular patch to be obliterated, and rays substituted instead,—it is a tail set on awry. And in this fish also we find the ichthyic displacement of fore limb. One step more, and we arrive at the homocercal or equal-lobed tail, which seems to attain to its most extreme type in those fishes in which, as in the perch and flounder, the last vertebral joint, either very little or very abruptly diminished in size, expands into broad processes, without homologue in the higher animals, on which the caudal rays rest as their bases. And in by much the larger proportion of these fishes all the four limbs are slung round the neck ;—they at once exhibit the homocercal tail in its broadest type, and displacement of limb in its most extreme form.

Now, in tracing the geologic history of the ichthyic tail, we find these several steps or gradations from the heterocercal to the homocercal, represented by periods and formations. The Silurian periods may be regarded as representative of that true heterocercal tail of the placoids, exemplified in *Spinax* (page 148, fig. 48), and *Cestracion* (page 153, fig. 49). The whole caudal portion of this latter animal, commencing immediately behind the ventrals, is, as becomes a true tail, slim, when compared with its trunk ; the vertebræ are of very considerable solidity ; the rays mucoidal; and where the spinal column runs into the terminal fin, it takes such an upward turn as that which the horse-jockey imparts, by the process of *nicking*, to the tails of the hunter and the race-horse. And with the heterocercal tail, so true in its homologies to the tails of the higher vertebrata, we find associated, as has been shown, the true homological position of the fore limbs. With the commencement of the Old Red Sandstone the ganoidal tail first presents itself ; and we become sensible of a change in the structure of the attached fin, similar to that exemplified in the caudal rays of the sturgeon. As shown by the irregularly-angular patch of scales which in all the true Celacanths, and almost all the Dipterians,* runs

* The vertebral column in the genus *Diplopterus* ran, as in the placoid genus *Scyllium*, nearly through the middle of the caudal fin.

through the *upper* lobe of the fin, and terminates in a point (see fig. 50), it must have possessed the gradually-diminish-

Fig. 50.

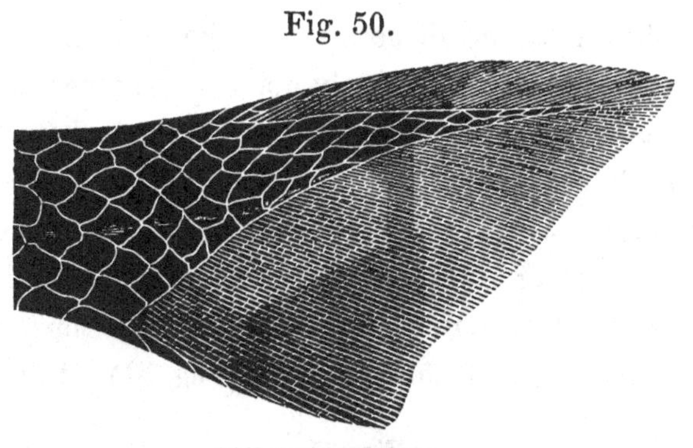

TAIL OF OSTEOLEPIS.

ing vertebræ, or a diminishing spinal cord, their analogue; but the rays, fairly *set*, as their state of keeping in the rocks certify, exist as narrow oblong plates of solid bone; and their anterior edges are strengthened by a line of osseous defences, that pass from scales into rays. And in harmonious accompaniment with this fairly *stereotyped* edition of the ichthyic tail, we find, in the fishes in which it appears, the first instance of displacement of *limb*,—the bases of the pectorals being removed from their original position, and stuck on to the nape of the neck. It may be remarked, in passing, that in the tails of two ganoidal genera of this period,—the *Coccosteus* and *Pterichthys*,—the analogies traceable lie rather in the direction of the tails of the Rays than in those of the Sharks; and that one of these, the *Coccosteus*, seems, as has been already intimated, to have had no pectorals, while it is doubtful whether in the *Pterichthys* the pectorals were not attached to the shoulder, instead of the head. In the Carboniferous and Permian systems there occur, especially among the numerous species of the genus *Palæoniscus*, tails of the type exemplified by the internal angle of the tail of the sturgeon: the lozenge-shaped scales run in acutely angular patches through their upper lobes; but such is their extreme flatness, as shown by the disposition of the enamelled covering, that it appears exceedingly doubtful whether any vertebral column ran beneath;—they seem but to have covered greatly diminished prolongations of the spinal cord. In the base of the Secondary division,—another long stage towards the existing state of things,—we find, with the homocercal tail, which now appears for the first time, numerous tails like that of the *Lepidosteus* (fig. 51), of an inter-

Fig. 51.

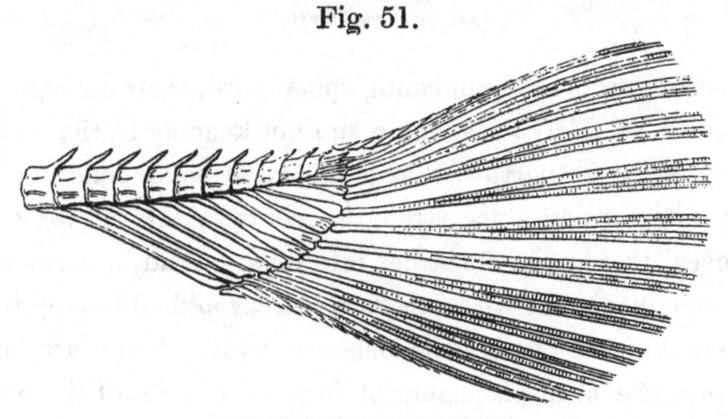

TAIL OF LEPIDOSTEUS OSSEUS

mediate type;—they are rather tails set on awry than truly heterocercal. The diminished cord has disappeared from among the fin rays. In the numerous Lepidoid genus, and the genera *Semionotus* and *Tetragonolepis*,—all ganoidal fishes of the Secondary period,—this intermediate style is very marked; while in their contemporaries of the genera *Uræus, Microdon*, and *Pycnodus*, we find the earliest examples of true homocercal tails. And in the ctenoids and cycloids of the Chalk the homocercal tail receives its fullest development. It finds bases for its rays in broad non-homological processes, that spread out behind abruptly-terminating vertebræ (fig. 52), in

Fig. 52.

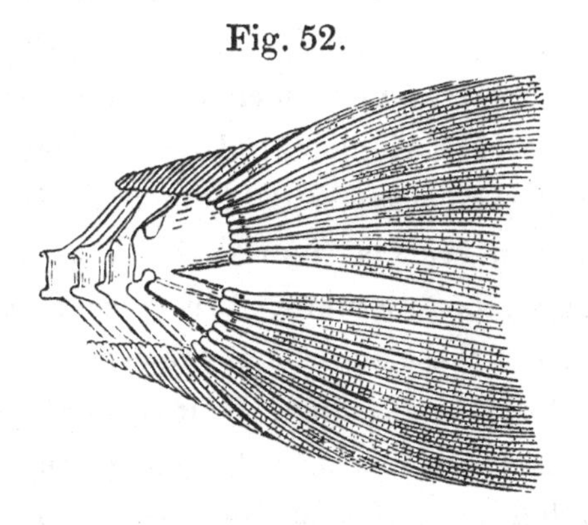

TAIL OF PERCH.

the same period in which, by a strange process of degradation, the four ichthyic limbs are first gathered into a cluster, and hung about the neck.

I am aware that by some very distinguished comparative anatomists, among the rest Professor Owen, the attachment, so common among fishes, of the scapular arch and the fore limbs to the occipital bone, is regarded, not as a displacement, but as a normal and primary condition of the parts. Recognising in the scapular bones the *ribs* of the occipital *centrum*, the anatomists of this school of course consider them, when found articulated to the occiput, as in their proper and original place, and as in a state of natural dislocation when removed, as in all the reptiles, birds, and mammals, farther down. We find Professor Oken borrowing support to his hypothesis from this view. The limbs, he tells us, are simply ribs that, in the course of ages, have been set free, and have become by development what they now are. And it is unquestionably a curious and interesting fact, that there are certain animals, such as the crocodile, in which every centrum of the vertebral column, and of every *vertebra* of the head, has its ribs, or rib-like appendages, with the exception of the occipital *centrum*. And it is another equally curious fact, that there is another certain class of animals, such as the osseous horn-covered fishes, with the Sturionidæ, Salamandroidei, and at least one genus among the placoids (the Chimæroidei), in which this occipital centrum bears as its *ribs* the scapular bones, with their appendages the fore limbs. It is the *centrum* without *ribs* that is selected in these animals as the centrum to which the scapular *ribs* should be attached. Be it remembered, however, that while it is unquestionably the part of the comparative anatomist to determine the relations and homologies of those parts of which all animals are composed, and to interpret the significancy in the scale of being of the various modes and forms in which they exist, it is as unquestionably the part of the geologist to declare their history, and the order of their succession *in time*. The questions which fall to be determined by the geologist and anatomist are entirely different. It is the function of the anatomist to decide regarding the high and the low, the typical and the aberrant ; and so, beginning at what is lowest or highest in the scale, or least or most symmetrical in type, he passes through the intermediate forms to the opposite extreme : and such is the order natural and proper to his science. It is the vocation of the geologist, on the other hand, to decide regarding the early and the late. It is with *time*, not with *rank*, that he has to deal. Nor is it in the least surprising that he should seem at issue with the comparative anatomist, when, in classifying his groupes of organized being according to the periods of their appearance, there is an order of arrangement forced upon him, different from that which, on an entirely different principle, the anatomist pursues. Nor can there be a better illustration of a collision of this kind, than the one furnished by the case in point. That peculiarity of structure which, as the lowest in the vertebral skeleton, is to the comparative anatomist the primary and original one, and which, as such, furnishes him with his starting point, is to the geologist not primary, but secondary, simply because it was not primary, but secondary, in the order of its occurrence. It belongs, so far as we yet know, not to the *first* period of vertebrate existence, but to the *second*; and appears in geologic history as does that savage state which certain philosophers have deemed the original condition of the human species, in the history of civilization, when read by the light of the Revealed Record, under the shadow of those gigantic ruins of the East that date only a few centuries after the Flood. It is found to be a *degradation* first introduced during the lapse of an intermediate age,—not the normal condition which obtained during the long cycles of the primal one. It indicates, not the starting point from which the race of creation began, but the stage of retrogradation beyond it at which the pilgrims who set out in a direction opposite to that of the goal first arrived.*

---

* I would, however, respectfully suggest, that that theory of cerebral vertebræ, on which, in this question, the comparative anatomists proceed as their principle, and which finds as little support in the geologic record from the actual history of the fore limbs as from the actual history of the bones of the cranium, may be more ingenious than sound. It is a shrewd circumstance, that the rocks refuse to testify in its favour. Agassiz, I find, decides against it on other than geological grounds; and his conclusion is certainly rendered not the less worthy of careful consideration by the fact that, yielding to the force of evidence, his views on the subject underwent a thorough change. He had first held, and then rejected it. "I have shared," he says, "with a multitude of

This fact of degradation, strangely indicated in geologic history, with reference to all the greater divisions of the animal kingdom, has often appeared to me a surpassingly

---

other naturalists, the opinion which regards the cranium as composed of vertebræ; and I am consequently in some degree called upon to point out the motives which have induced me to reject it."

"M. Oken," he continues, "was the first to assign this signification to the bones of the cranium. The new doctrine he expounded was received in Germany with great enthusiasm by the school of the philosophers of nature. The author conceived the cranium to consist of three [four] vertebræ, and the basal occipital, the sphenoid, and the ethmoid, were regarded as the central parts of these cranial vertebræ. On these alleged bodies of vertebræ, the arches enveloping the central parts of the nervous system were raised, while on the opposite side were attached the inferior pieces, which went to form the vegetative arch destined to embrace the intestinal canal and the large vessels. It would be too tedious to enumerate in this place the changes which each author introduced, in order to modify this matter so as to make it suit his own views. Some went the length of affirming that the vertebræ of the head were as complete as those of the trunk; and, by means of various dismemberments, separations, and combinations, all the forms of the cranium were referred to the vertebræ, by admitting that the number of pieces was invariably fixed in every head, and that all the vertebrata, whatever might be their organization in other respects, had in their heads the same number of points of ossification. At a later period, what was erroneous in this manner of regarding the subject was detected; but the idea of the vertebral composition of the head was still retained. It was admitted as a general law, that the cranium was composed of three primitive vertebræ, as the embryo is of three blastodermic leaflets; but that these vertebræ, like the leaflets, existed only ideally, and that their presence, although easily demonstrated in certain cases, could only be slightly traced, and with the greatest difficulty, in other instances. The notion thus laid down of the virtual existence of cranial vertebræ did not encounter very great opposition; it could not be denied that there was a certain general resemblance between the osseous case of

wonderful one. We can see but imperfectly, in those twilight depths to which all such subjects necessarily belong; and yet at times enough does appear to show us what a very

---

the brain and the rachidian canal; the occipital, in particular, had all the characteristic features of a vertebra. But whenever an attempt was made to push the analogy further, and to determine rigorously the anterior vertebræ of the cranium, the observer found himself arrested by insurmountable obstacles, and he was obliged always to revert to the virtual existence.

"In order to explain my idea clearly, let me have recourse to an example. It is certain that organized bodies are sometimes endowed with virtual qualities, which, at a certain period of the being's life, elude dissection, and all our means of investigation. It is thus that, at the moment of their origin, the eggs of all animals have such a resemblance to each other, that it would be impossible to distinguish, even by the aid of the most powerful microscope, the ovarial egg of a craw-fish, for example, from that of true fish. And yet who would deny that beings in every respect different from each other exist in these eggs? It is precisely because the difference manifests itself at a later period, in proportion as the embryo develops itself, that we were authorized to conclude, that, even from the earliest period, the eggs were different,—that each had virtual qualities proper to itself, although they could not be discovered by our senses. If, on the contrary, any one should find two eggs perfectly alike, and should observe two beings perfectly identical issue from them, he would greatly err if he ascribed to these eggs different virtual qualities. It is therefore necessary, in order to be in a condition to suppose that virtual properties peculiar to it are concealed in an animal, that these properties should manifest themselves once, in some phase or other of its development. Now, applying this principle to the theory of cranial vertebræ, we should say, that if these vertebræ virtually exist in the adult, they must needs show themselves in reality, at a certain period of development. If, on the contrary, they are found neither in the embryo nor the adult, I am of opinion that we are entitled likewise to dispute their virtual existence.

"Here, however, an objection may be made to me, drawn from

superficial thing infidelity may be. The general advance in creation has been incalculably great. The lower divisions of the vertebrata preceded the higher ;—the fish preceded the

---

the physiological value of the vertebræ, the function of which, as is well known, is, on the one hand, to furnish a solid support to the muscular contractions which determine the movements of the trunk, and, on the other, to protect the centres of the nervous system, by forming a more or less solid case completely around them. The bodies of the vertebræ are particularly destined to the first of these offices; the neurapophyses to the second. What can be more natural than to admit, from the consideration of this, that in the head, the bodies of the vertebræ diminish in proportion as the moving function becomes lost, while the neurapophyses are considerably developed for protecting the brain, the volume of which is very considerable, when compared with that of the spinal marrow? Have we not an example of this fact in the vertebræ of the tail, where the neurapophyses become completely obliterated, and a simple cylindrical body alone remains? Now, may it not be the case, that in the head, the bodies of the vertebræ have disappeared; and that, in consequence, there is a prolongation of the cord only as far as the moving functions of the vertebræ extend? There is some truth in this argument, and it would be difficult to refute it *a priori*. But it loses all its force the moment that we enter upon a detailed examination of the bones of the head. Thus, what would we call, according to this hypothesis, the principal sphenoid, the great wings of the sphenoid, and the ethmoid, which form the floor of the cerebral cavity? It may be said they are apophyses. But the apophyses protect the nervous centres only on the side and above. It may be said that they are the bodies of the vertebræ. But they are formed without the concurrence of the dorsal cord; they cannot, therefore, be the bodies of the vertebræ. It must therefore be allowed, that these bones at least do not enter into the vertebral type; that they are in some measure peculiar. And if this be the case with them, why may not the other protective plates be equally independent of the vertebral type; the more so, because the relations of the frontals and parietals vary so much, that it would be almost impossible to assign to them a constant place?"

reptile, the reptile preceded the bird, the bird preceded the mammiferous quadruped, and the mammiferous quadruped preceded man. And yet, is there one of these great divisions in which, in at least some prominent feature, the present, through this mysterious element of degradation, is not inferior to the past? There was a time in which the ichthyic form constituted the highest example of life; but the seas during that period did not swarm with fish of the degraded type. There was, in like manner, a time when all the carnivora and all the herbivorous quadrupeds were represented by reptiles; but there are no such magnificent reptiles on the earth now as reigned over it then. There was an after time, when birds seem to have been the sole representatives of the warm-blooded animals; but we find, from the prints of their feet left in sandstone, that the tallest men might have

> "Walked under their huge legs, and peeped about."

Further, there was an age when the quadrupedal mammals were the magnates of creation; but it was an age in which the sagacious elephant, now extinct, save in the comparatively small Asiatic and African circles, and restricted to two species, was the inhabitant of every country of the Old World, from its southern extremity to the frozen shores of the northern ocean; and when vast herds of a closely allied and equally colossal genus occupied its place in the New. And now, in the times of the high-placed human dynasty,—of those formally delegated monarchs of creation, whose nature it is to look behind them upon the past, and before them, with mingled fear and hope, upon the future,—do we not as certainly see the elements of a state of ever-sinking degradation, which is to exist for ever, as of a state of ever-increasing perfectibility, to which there is to be no end? Nay, of a higher race, of which we know but little, this much we at least know, that they long since separated into two great classes,—that of the "elect angels," and of "angels that kept not their first estate."

EVIDENCE OF THE SILURIAN MOLLUSCS.—OF THE FOSSIL FLORA.

ANCIENT TREE.

AFTER dwelling at such length on the earlier fishes, it may seem scarce necessary to advert to their lower contemporaries the mollusca,—that great division of the animal kingdom which Cuvier places second in the descending order, in his survey of the entire series, and first among the invertebrates; and which Oken regards as the division out of which the immediately preceding class of the vertebral animals have been developed. "The fish," he says, "is to be viewed as a mussel, from between whose shells a monstrous abdomen has grown out." There is, however, a peculiarity in the molluscan group of the Silurian system, to which I must be permitted briefly to refer, as, to employ the figure of Sterne, it presents "two handles" of an essentially different kind, and as in all such two-handled cases, the mere special pleader is sure to avail himself of only the handle which best suits his purpose for the time.

Cuvier's first and highest class of the mollusca is formed of what are termed the Cephalopods,—a class of creatures possessed of great freedom of motion: they can walk, swim, and seize their prey; they have what even the lowest fishes, such as the lancelet, want,—a brain enclosed in a cartilaginous cavity in the head, and perfectly formed organs of sight; they possess, too, what is found in no other mollusc,—organs of hearing; and in sagacity and activity they prove more than matches for the smaller fishes, many of which they overmaster and devour. With this highest class there contrasts an exceedingly low molluscous class at the bottom of the scale, or, at least, at what is now the bottom of the scale; for they constitute Cuvier's *fifth* class; while his *sixth* and last, the Cirrhopodes, has been since withdrawn from the molluscs altogether, and placed in a different division of the animal kingdom. And this low class, the Brachipods, are creatures that, living in bivalve shells, unfurnished with spring hinges to throw them open, and always fast anchored to the same spot, can but thrust forth, through the interstitial chinks of their prison-houses, spiral arms, covered with cilia, and winnow the water for a living. Now, it so happens that the molluscan group of the Silurian system is composed chiefly of these two extreme classes. It contains some of the other forms; but they are few in number, and give no character to the rocks in which they occur. There was nothing by which I was more impressed, in a visit to a Silurian region, than that in its ancient grave-yards, as in those of the present day, though in a different sense, the high and the low should so invariably meet together. It is, however, not impossible that, in even the present state of things, a similar union of the extreme forms of the marine mollusca may be taking place in deep-sea deposits. Most of the intermediate forms provided with shells capable of preservation, such as the shelled Gasteropoda and the Conchifers, are either littoral, or restricted to comparatively small depths; whereas the Brachipoda are deep-sea shells; and the Cephalopoda may be found voyaging far from land, in the upper strata of the sea above them. Even in the seas that surround our own island, the Brachipodous molluscs,—terebratula and crania,—have been found, ever since deep-sea dredging became common, to be not very rare shells; and in the Mediterranean, where they are less rare still, fleets of Argonauts, the representatives of a highly organized family of the Cephalopods, to which it is now believed the Bellerophon of the Palæozoic rocks belonged, may be seen skimming along the surface, with sail and oar, high over the profound depths in which they lie. And, of course, when death comes, that comes to high and low, the remains of both Argonauts and Brachipods must lie together at the bottom, in beds almost totally devoid of the intermediate forms.

Now, the author of the "Vestiges," in maintaining his hypothesis, suspends it on the handle furnished him by the immense abundance of the Silurian Brachipods. The Silurian period, he says, exhibits "a scanty and most defective development of life; so much so, that Mr Lyell calls it, *par excellence*, the age of Brachipods, with reference to the by no means exalted bivalve shell-fish which forms its predominant class. Such being the actual state of the case, I must persist in describing even the fauna of this age, which we now know was not the first, as, generally speaking, such a humble exhibition of the animal kingdom as we might expect, upon the development theory, to find at an early stage of the history of organization." The reader will at once discern the fallacy here. The Silurian period was peculiarly an age of Brachipods, for in no other period were Brachipods so numerous, specifically or individually, or of such size or importance; whereas it was not *so peculiarly* an age of Cephalopods, for these we find introduced in still greater numbers during the Liasic and Oolitic periods. In 1848, when

Professor Edward Forbes edited the Palæontological map of Britain and Ireland, which forms one of the very admirable series of "Johnstone's Physical Atlas," the Cephalopods of the Silurian rocks of England and Wales were estimated at forty-eight species, and the Brachipods at one hundred and fifty; whereas at the same date there were two hundred and five Cephalopods of the Oolitic formations enumerated, and but fifty-four Brachipods. It is the molluscs of the inferior, not those of the superior class, that constitute (with their contemporaries the Trilobites) the characteristic fossils of the Silurian rocks; and hence the propriety of the distinctive name suggested by Sir Charles Lyell. But in the development question, what we have specially to consider is, not the *numbers* of the low, but the *standing* of the high. A country may be distinctively a country of flocks and herds, or a country of the carnivorous mammalia, or, like New South Wales or the Galapagos, a country of marsupial animals or of reptiles. Its human inhabitants may be merely a few hunters or shepherds, too inconsiderable in numbers, and too much like their brethren elsewhere, to give it any peculiar standing as a home of men. But in estimating the highest point in the scale to which the animal kingdom has attained within its limits, it is of its few men, not of its many beasts, that we must take note. And the point to be specially decided regarding the organisms of the Silurian system, in this question, is, not the proportion in *number* which the lower forms bore to the higher, but the exact *rank* which the higher bore in the scale of existence. Did the system furnish but a single Cephalopod or a single fish, we would yet have as certainly to determine that the chain of being reached as high as the Cephalopod or the fish, as if the remains of these creatures constituted its most abundant fossils. The chain of animal life reached quite as high on the evening of the sixth day of creation, when the human family was restricted to a single pair, as it does now, when our statists reckon up by millions the inhabitants of the greater capitals of the world; and the special pleader who, in asserting the contrary, would insist on determining the point, not by the *rank* of the men of Eden, but by the *number* of minnows or sticklebacks that swarmed in its rivers, might be perhaps deemed ingenious in his expedients, but certainly not very judicious in the use of them. It is worthy of remark, however, that the Brachipods of those Palæozoic periods in which the group occupied such large space in creation, consisted of greatly larger and more important animals than any which it contains in the present day. It has yielded to what geological history shows to be the common fate, and sunk into a state of degradation and decline.

The geological history of the vegetable, like that of the animal kingdom, has been pressed into the service of the development hypothesis; and certainly their respective courses, both in actual arrangement and in their relation to human knowledge, seem wonderfully alike. It is not much more than twenty years since it was held that no exogenous plant existed during the Carboniferous period. The frequent occurrence of Coniferæ in the Secondary deposits had been conclusively determined from numerous specimens; but, founding on what seemed a large amount of negative evidence, it was concluded that, previous to the Liasic age, nature had failed to achieve a tree, and that the rich vegetation of the Coal Measures had been exclusively composed of magnificent immaturities of the vegetable kingdom,— of gigantic ferns and club-mosses, that attained to the size of forest-trees, and of thickets of the swamp-loving horsetail family of plants, that well nigh rivalled in height those forests of masts which darken the rivers of our great commercial cities. Such was the view promulgated by M. Adolphe Brogniart; and it may be well to remark that, so far as the evidence on which it was based was positive, the view was sound. It *is* a fact, that inferior orders of plants were developed in those ages in a style which in their present state of degradation they never exemplify: they took their place, not, as now, among the pigmies and abortions of creation, but among its tallest and goodliest productions. It is, however, *not* a fact that they were the highest vegetable forms of their time. True exogenous trees also existed in great numbers and of vast size. In various localities in the coal fields of both England and Scotland,—such as Lennel Braes and Allan Bank in Berwickshire, High-Heworth, Fellon, Gateshead, and Wideopen near Newcastle-upon-Tyne, and in quarries to the west of the city of Durham,—the most abundant fossils of the system are its true woods. In the quarry of Craigleith, near Edinburgh, three huge trunks have been laid open during the last twenty years, within the space of about a hundred and fifty yards, and two equally massy trunks, within half that space, in the neighbouring quarry of Granton,—all low in the Coal Measures. They lie diagonally athwart the strata,—at an angle of about thirty,—with the nether and weightier portion of their boles below, like snags in the Mississippi; and we infer, from their general direction, that the stream to which they reclined must have flowed from nearly north-east to south-west. The current was probably that of a noble river, which reflected on its broad bosom the shadow of many a stately tree. With the exception of one of the Granton specimens, which still retains its strong-kneed roots, they are all mere portions of trees, rounded at both ends, as if by attrition or decay; and yet one of these portions measures about six feet in diameter by sixty-one feet in length; another four feet in diameter by seventy feet in length; and the others, of various thickness, but all bulky enough to equal the masts of large vessels, range in length from thirty-six to forty-seven feet. It seems strange to one who derives his supply of domestic fuel from the Dalkeith and Falkirk coal-fields, that the Carboniferous flora could ever have been described as devoid of trees. I can scarce take up a piece of coal from beside my study fire without detecting in it fragments of carbonized wood, which almost always exhibit the characteristic longitudinal fibres, and not unfrequently the medullary rays. Even the trap-rocks of the district enclose, in some instances, their masses of lignite, which present in their transverse sections, when cut by the lapidary, the net-like reticulations of the coniferæ. The fossil botanist who devoted himself chiefly to the study of microscopic structure would have to decide, from the facts of the case, not that trees were absent during the Carboniferous period, but that, in consequence of their having been present in amazing numbers, their remains had entered more palpably and extensively into the composition of coal than those of any other vegetable.* So far as is yet known, they all belonged to the

* It is stated by Mr Witham, that, "except in a few instances, he had ineffectually tried, with the aid of the microscope, to obtain some insight into the structure of coal. Owing," he adds, "to its great opacity, which is probably due to mechanical pressure, the action of chemical affinity, and the percolation of acidulous waters, all traces of organization appear to have been obliterated." I have heard the late Mr Sanderson, who prepared for Mr Witham most of the specimens figured in his well-known work on the "Internal Structure of Fossil Vegetables," and from whom the

two great divisions of the coniferous family, araucarians and pines. The huge trees of Craigleith and Granton were of the former tribe, and approximate more nearly to *Altingia excelsa,*

Fig. 53.

ALTINGIA EXCELSA (NORFOLK-ISLAND PINE.)
From a young specimen in the Botanic Garden, Edinburgh.

---

materials of his statement on this point seem to have been derived, make a similar remark. It was rare, he said, to find a bit of coal that exhibited the organic structure. The case, however, is far otherwise; and the ingenious mechanic and his employer were misled, simply by the circumstance, that it is rare to find pieces of coal which exhibit the ligneous fibre, existing in a state of keeping solid enough to stand the grinding of the lapidary's wheel. The lignite usually occurs in thin layers of a substance resembling soft charcoal, at which, from the loose adhesion of the fibres, the coal splits at a stroke; and as it cannot be prepared as a transparency, it is best examined by a Stanhope lens. It will be found, tried in this manner, that so far is vegetable fibre from being of rare occurrence in coal,—our Scotch coal at least,—that almost every cubic inch contains its hundreds, nay, its thousands, of cells.

the Norfolk-Island pine,—a noble araucarian, that rears its proud head from a hundred and sixty to two hundred feet over the soil, and exhibits a green and luxuriant breadth of foliage rare among the Coniferæ,—than any other living tree.

Beyond the Coal Measures terrestrial plants become extremely rare. The fossil botanist, on taking leave of the lower Carboniferous beds, quits the land, and sets out to sea; and it seems in no way surprising, that the specimens which he there adds to his herbarium should consist mainly of *Fucaceæ* and *Conferveæ*. The development hypothesis can borrow no support from the simple fact, that while a high terrestrial vegetation grows upon dry land, only algæ grow in the sea; and even did the Old Red Sandstone and Silurian systems furnish, as their vegetable organisms, fucoids exclusively, the evidence would amount to no more than simply this, that the land of the Palæozoic periods produced plants of the land, and the sea of the Palæozoic periods produced plants of the sea.

In the Upper Old Red Sandstone,—the formation of the *Holoptychius* and the *Stagonolepis*,—the only vegetable remains which I have yet seen are of a character so exceedingly obscure and doubtful, that all I could venture to premise regarding them is, that they *seem* to be the fragments of sorely comminuted fucoids. In the formation of the Middle Old Red,—that of the Cephalaspis and the gigantic lobster of Carmylie,—the vegetable remains are at once more numerous and better defined. I have detected among the gray micaceous sandstones of Forfarshire a fucoid furnished with a thick, squat stem, that branches into numerous divergent leaflets or fronds, of a slim parallelogrammical, grass-like form, and which, as a whole, somewhat resembles the scourge of cords attached to a handle with which a boy whips his top. And

Professor Fleming describes a still more remarkable vegetable organism of the same formation, " which, occurring in the form of circular, flat patches, composed each of numerous smaller contiguous circular pieces, is altogether not unlike what might be expected to result from a compressed berry, such as the bramble or rasp." In the Lower Old Red,—the formation of the *Coccosteus* and *Cheiracanthus*,—the remains of fucoids are more numerous still. There are gray slaty beds among the rocks of Navity, that owe their fissile character mainly to their layers of carbonized weed ; and " among the rocks of Sandy-Bay, near Thurso," says Mr Dick, " the dark impressions of large fucoids are so numerous, that they remind one of the interlaced boughs and less bulky pine-trunks that lie deep in our mosses." A portion of a stem from the last locality, which I owe to Mr Dick, measures three inches in diameter ; but the ill-compacted cellular tissue of the algæ is but indifferently suited for preservation ; and so it exists as a mere coaly film, scarcely half a line in thickness.

The most considerable collection of the Lower Old Red fucoids which I have yet seen is that of the Rev. Charles Clouston of Sandwick, in Orkney,—a skilful cultivator of geological science, who has specially directed his palæontological inquiries on the vegetable remains of the flagstones of his district, as the department in which most remained to be done ; but his numerous specimens only serve to show what a poverty-stricken flora that of the ocean of the Lower Old Red Sandstone must have been. I could detect among them but two species of plants ;—the one an imperfectly preserved vegetable, more nearly resembling a club-moss than aught else which I have seen, but which bore on its surface, instead of the well-marked scales of the *Lycopodiaceæ*, irregular rows of tubercles, that, when elongated in the profile, as sometimes happens, might be mistaken for minute, ill-defined leaves ; the other, a smooth-stemmed fucoid, existing on the stone in most cases as a mere film, in which, however, thickly-set longitudinal fibres are occasionally traceable, and which may be always distinguished from the other by its sharp-edged outline, and from the circumstance that its stems continue to retain the same diameter for considerable distances, after throwing off at acute angles numerous branches nearly as bulky as themselves. In a Thurso specimen, about two feet in length, which I owe to the kindness of Mr Dick, there are stems continuous throughout, that, though they ramify in that space into from six to eight branches, are nearly as thick atop as at bottom. They are the remains, in all probability, of a long, flexible weed, that may have somewhat resembled those fucoids of the intertropical seas which, streaming slantwise in the tide, rise not unfrequently to the surface in from fifteen to twenty fathoms water; and as, notwithstanding their obscurity, they are among the most perfect specimens of their class yet found, and contrast with the stately araucarians of the Coal Measures, in a style which cannot fail to delight the heart of every assertor of the development hypothesis, I present them to the reader from Mr Dick's specimen, in a figure (fig. 54) which, however slight its interest, has at least the merit of being true. The stone exhibits specimens of the two species of Mr Clouston's collection,—the sharp-edged, finely-striated weed, *a*, and that roughened by tubercles, *b*; which, besides the distinctive character manifested on its surface, differs from the other in rapidly losing breadth with every branch which it throws off, and, in consequence, runs soon to a point. The cut on the opposite page (fig. 55) represents not inadequately the cortical peculiarities of the two

Fig. 54.

FUCOIDS OF THE LOWER OLD RED SANDSTONE.

a. *Smooth-stemmed species.* b. *Tubercled species.*

(One-sixth nat. size, linear.)

Fig. 55.

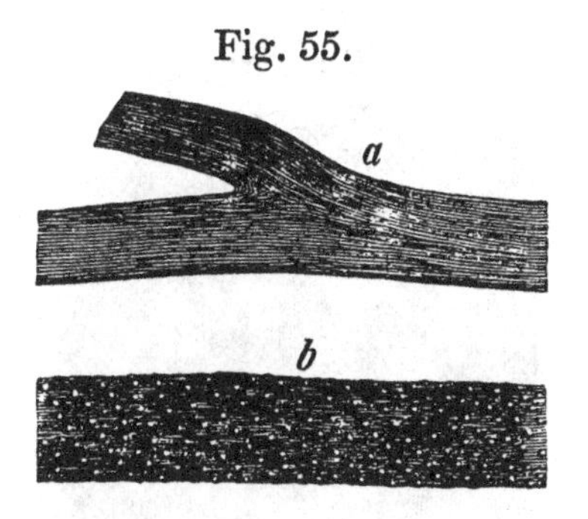

a. *Smooth-stemmed species.* b. *Tubercled species.*

(Natural size.)

species when best preserved. The surface of the tubercled one will perhaps remind the Algologist of the knobbed surface of the thong or receptacle of *Himanthalia lorea*, a recent fucoid, common on the western coast of Scotland, but rare on the east. An Orkney specimen lately sent me by Mr William Watt, from a quarry at Skaill, has much the appearance of one of the smaller ferns, such as the moor-worts, sea spleen-worts, or maiden-hairs. It exists as an impression in diluted black, on a ground of dark gray, and has so little sharpness of outline, that, like minute figures in oil-paintings, it seems more distinct when viewed at arm's length than when microscopically examined; but enough remains to show that it must have been a terrestrial, not a marine plant. The accompanying print (fig. 56) may be regarded as no unfaithful representation of this unique fossil in its state of imperfect keeping. The vegetation of the Silurian system, from its upper beds down till where we reach the zero of life, is, like that of the Old Red Sandstone, almost exclusively fucoidal. In the older fossiliferous deposits of the system in Sweden, Russia, the Lake Districts of England, Canada, and the United States, fucoids occur, to the exclusion, so far as is

Fig. 56.

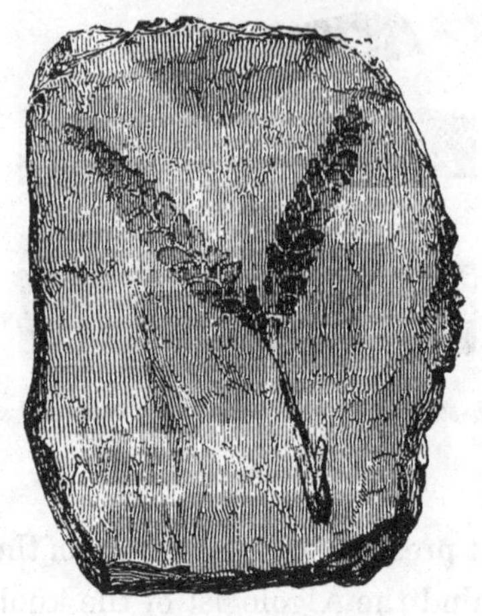

FERN ? OF THE LOWER OLD RED SANDSTONE.

(Natural size.)

yet known, of every other vegetable form ; and such is their abundance in some localities, that they render the argillaceous rocks in which they lie diffused, capable of being fired as an alum slate, and exist in others as seams of a compact anthracite, occasionally used as fuel. They also occur in those districts of Wales in which the place and sequence of the various Silurian formations were first determined, though apparently in a state of keeping from which little can be premised regarding their original forms. Sir Roderick Murchison sums up his notice of the vegetable remains of the system in the province whence it derives its name, by stating, that he had submitted his specimens to "Mr Robert Brown and Dr Greville, and that neither of these eminent botanists were able to say much more regarding them than that they were fucoid-like bodies."

Such are the vegetable organisms of the Old Red Sandstone and Silurian systems : they are the remains of the ancient marine plants of ancient marine deposits, and, as such, lend quite as little support to the development hypothesis as the recent algæ of our existing seas. The case, stated in its most favourable form, amounts simply to this,—that at certain early periods,—represented by the Upper and Lower Silurian and the Old Red deposits,—the seas produced sea-plants; and that at a certain later period,—that of the Carboniferous system,—the land produced land-plants. But even this, did it stand alone, would be a *too* favourable statement. I have seen, on one occasion, the fisherman bring up with his nets, far in the open sea, a wild rose-bush, that, though it still bore its characteristic thorns, was encrusted with serpula, and laden with pendulous lobularia. It had been swept from its original habitat by some river in flood, that had undermined and torn down the bank on which it grew; and after floating about, mayhap for months, had become so saturated with water, that it could float no longer. And in that single rose-bush, dragged up to the light and air from its place among Sertularia, Flustra, Serpula, and the deep-sea fucoids, I had as certain an evidence of the existence of the dicotyledonous plant, as if I had all the families of the Rosaceæ before me. Now, we are furnished by the more ancient formations with evidence regarding the existence of a terrestrial vegetation, such as that which the rose-bush in this case supplied. We cannot expect that the proofs should be numerous. In the chart of the Pacific attached to the better editions of "Cook's Voyages," there are several notes along the tract of the great navigator, that indicate where, in mid ocean, trees or fragments of trees had been picked up. These entries, however, are but few, though they belong to all the three voyages together: if I remember aright, there are only five entries in all,—two in the Northern, and three in the Southern Pacific. The floating shrub or tree, at a great distance from land, is of rare occurrence in even the present scene of things, though the breadth of land be great, and trees numerous ; and in the times of the Silurian and Old Red Sandstone systems, when the breadth of land was apparently *not* great, and trees and shrubs, in consequence, *not* numerous, it must have been of rarer occurrence still. We learn, however, from Sir Charles Lyell, that in the "Hamilton group of the United States,—a series of beds that corresponds in many of its fossils with the Ludlow rocks of England,—plants allied to the *Lepidodendra* of the Carboniferous type are abundant ; and that in the lower Devonian strata of New York the same plants occur associated with ferns." And I am able to demonstrate, from an interesting fossil at present before me, that there existed in the period of the Lower Old Red Sandstone, vegetable forms of a class greatly higher than either *Lepidodendra* or ferns.

In my little work on the Old Red Sandstone, I have referred to an apparent lignite of the Lower Old Red of Cromarty, which presented, when viewed by the microscope, marks of the internal fibre. The surface, when under the glass, resembled, I said, a bundle of horse-hairs lying stretched in parallel lines : and in this specimen alone, it was added, had I found aught in the Lower Old Red Sandstone approaching to proof of the existence of dry land. About four years ago I had this lignite put stringently to the question by Mr Sanderson ; and deeply interesting was the result. I must first mention, however, that there cannot rest the shadow of a doubt regarding the place of the organism in the geologic scale. It is unequivocally a fossil of the Lower Old Red Sandstone. I found it partially embedded, with many other nodules half-disinterred by the sea, in an ichthyolitic deposit, a few hundred yards to the east of the town of Cromarty, which occurs more than four hundred feet over the Great Conglomerate base of the system. A nodule that lay immediately beside it contained a well-preserved specimen of the *Coccosteus Decipiens*; and in the nodule in which the lignite itself is contained (fig. 57),

Fig. 57.

LIGNITE OF THE LOWER OLD RED SANDSTONE.

(One-third nat. size, linear.)

the practised eye may detect a scattered group of scales of *Diplacanthus*, a scarce less characteristic organism of the lower formation. And what, asks the reader, is the character of this very ancient vegetable,—the most ancient, by three whole formations, that has presented its internal structure to the microscope? Is it as low in the scale of development as in the geological scale? Does this venerable Adam of the forest appear, like the Adam of the infidel, as a squalid, ill-formed savage, with a rugged shaggy nature, which it would require the suggestive necessities of many ages painfully to lick into civilization? Or does it appear rather like the Adam of the poet and the theologian, independent, in its instantaneously-derived perfection, of all after development?

> "Adam, the goodliest man of men since born
> His sons."

Is its tissue vascular or cellular, or, like that of some of the cryptogamia, intermediate? Or what, in fine, is the nature and bearing of its mute but emphatic testimony, on that doctrine of progressive development of late so strangely resuscitated?

In the first place, then, this ancient fossil is a true wood,—a Dicotyledonous or Polycotyledonous *Gymnosperm*, that, like the pines and larches of our existing forests, bore naked seeds, which, in their state of germination, developed either double lobes to shelter the embryo within, or shot out a fringe of verticillate spikes, which performed the same protective functions, and that, as it increased in bulk year after year, received its accessions of growth in outside layers. In the transverse section the cells bear the reticulated appearance which distinguish the coniferæ (fig. 58, *a*); the lignite had been exposed in its bed to a considerable degree of pressure; and so the openings somewhat resemble the meshes of a net that has been drawn a little awry; but no general obliteration of their original character has taken place, save in minute patches, where they have been injured by compression or the bituminizing process. All the tubes indicated by the openings are, as in recent coniferæ, of nearly the same size; and though, as in many of the more ancient lignites, there are no indications of annual rings, the direction of the medullary rays is distinctly traceable. The longitudinal sections are rather less distinct than the transverse one: in the section parallel to the radius of the stem or bole the circular disks of the coniferæ

Fig. 58.

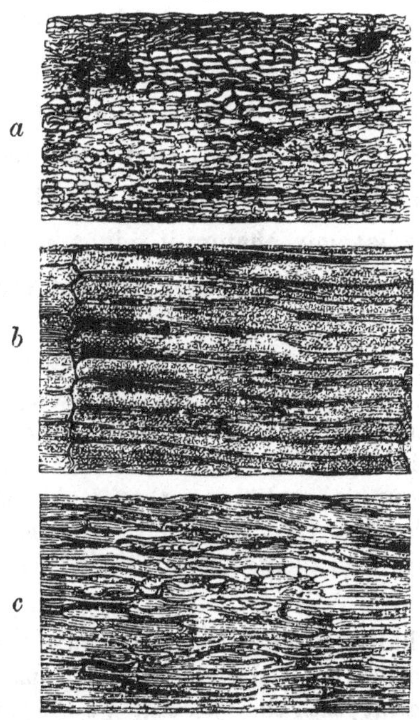

INTERNAL STRUCTURE OF LIGNITE OF LOWER OLD RED SANDSTONE.

a. *Transverse section.*
b. *Longitudinal section (parallel to radius, or medullary rays).*
c. *Longitudinal section (tangental, or parallel to the bark).*

(Mag. forty diameters.)

were at first not at all detected; and, as since shown by a very fine microscope, they appear simply as double and triple lines of undefined dots (*b*), that somewhat resemble the stippled markings of the miniature painter; nor are the openings of the medullary rays frequent in the tangental section (*i. e.* that parallel to the bark) (*c*); but nothing can be better defined than the peculiar arrangement of the woody fibre, and the longitudinal form of the cells. Such is the character of this, the most ancient of lignites yet found, that yields to the microscope the peculiarities of its original structure. We find in it an unfallen *Adam*,—not a half-developed savage.*

The olive leaf which the dove brought to Noah established at least three important facts, and indicated a few more. It showed most conclusively that there was dry land, that there were olive trees, and that the climate of the surrounding region, whatever change it might have undergone, was still favourable to the development of vegetable life.

---

* On a point of such importance I find it necessary to strengthen my testimony by auxiliary evidence. The following is the judgment, on this ancient petrifaction, of Mr Nicol of Edinburgh,—confessedly one of our highest living authorities in that division of fossil botany which takes cognizance of the internal structure of lignites, and decides, from their anatomy, their race and family:—

"Edinburgh, 19th July 1845.

"DEAR SIR,—I have examined the structure of the fossil wood which you found in the Old Red Sandstone at Cromarty, and have no hesitation in stating, that the reticulated texture of the transverse sections, though somewhat compressed, clearly indicates a coniferous origin; but as there is not the slightest trace of a disc to be seen in the longitudinal sections parallel to the medullary rays, it is impossible to say whether it belongs to the Pine or Araucarian division.—I am, &c.

"WILLIAM NICOL."

It will be seen that Mr Nicol failed to detect what I now deem the discs of this conifer,—those stippled markings to which I have referred, and which the engraver has indicated in no exaggerated style, in one of the longitudinal sections (*b*) of the wood-cut given above. But even were this portion of the evidence wholly wanting, we would be left in doubt, in consequence, not whether the Old Red lignite formed part of a true gymnospermous tree, but whether that tree is now represented by the pines of Europe and America, or by the araucarians of Chili and New Zealand. Were I to risk an opinion in a department not particularly my province, it would be in favour of an araucarian relationship.

And, farther, it might be very safely inferred from it, that if olive trees had survived, other trees and plants must have survived also ; and that the dark muddy prominences round which the ebbing currents were fast sweeping to lower levels, would soon present, as in antediluvian times, their coverings of cheerful green. The olive leaf spoke not of merely a partial, but of a general vegetation. Now, the coniferous lignite of the Lower Old Red Sandstone we find charged, like the olive leaf, with a various and singularly interesting evidence. It is something to know, that in the times of the *Coccosteus* and *Asterolepis* there existed dry land, and that that land wore, as at after periods, its soft, gay mantle of green. It is something also to know, that the verdant tint was not owing to a profuse development of the mere immaturities of the vegetable kingdom,—crisp, slow-growing lichens, or watery spore-propagated fungi that shoot up to their full size in a night,—nor even to an abundance of the more highly organized families of the liverworts and the mosses. These may have abounded then, as now ; though we have not a shadow of evidence that they did. But while we have no proof whatever of *their* existence, we have conclusive proof that there existed orders and families of a rank far above them. On the dry land of the Lower Old Red Sandstone, on which, according to the theory of Adolphe Brogniart, nothing higher than a lichen or a moss could have been expected, the ship-carpenter might have hopefully taken axe in hand, to explore the woods for some such stately pine as the one described by Milton,—

> " Hewn on Norwegian hills, to be the mast
> Of some great admiral."

Viewed simply in its picturesque aspect, this *olive leaf* of the Old Red seems not at all devoid of poetry. We sail upwards into the high geologic zones, passing from ancient to still more ancient scenes of being; and, as we voyage along, find ever in the surrounding prospect, as in the existing scene from which we set out, a graceful intermixture of land and water, continent, river, and sea. We first coast along the land of the Tertiary, inhabited by the strange quadrupeds of Cuvier, and waving with the reeds and palms of the Paris Basin; the land of the Wealden, with its gigantic iguanodon rustling amid its tree ferns and its cycadeæ, comes next; then comes the green land of the Oolite, with its little pouched, insectivorous quadruped, its flying reptiles, its vast jungles of the Brora equisetum, and its forests of the Helmsdale pine; and then, dimly as through a haze, we mark, as we speed on, the thinly scattered islands of the New Red Sandstone, and pick up in our course a large floating leaf, veined like that of a cabbage, which not a little puzzles the botanists of the expedition. And now we near the vast Carboniferous continent, and see along the undulating outline, between us and the sky, the strange forms of a vegetation, compared with which that of every previously seen land seems stunted and poor. We speed day after day along endless forests, in which gigantic club-mosses wave in air a hundred feet over head, and skirt interminable marshes, in which thickets of reeds overtop the mast-head. And, where mighty rivers come rolling to the sea, we mark, through the long-retiring vistas which they open into the interior, the higher grounds of the country covered with coniferous trees, and see doddered trunks of vast size, like those of Granton and Craigleith, reclining under the banks in deep muddy reaches, with their decaying tops turned adown the current. At length the furthermost promontory of this long range of coast comes full in view: we near it,—

we have come up abreast of it: we see the shells of the Mountain Limestone glittering white along its further shore, and the green depths under our keel lightened by the flush of innumerable corals; and then, bidding farewell to the land for ever,—for so the geologists of but five years ago would have advised,—we launch into the unmeasured ocean of the Old Red, with its three consecutive zones of animal life. Not a single patch of land more do those geologic charts exhibit which we still regard as new. The zones of the Silurian and Cambrian succeed the zones of the Old Red; and, darkly fringed by an obscure bank of cloud ranged along the last zone in the series, a night that never dissipates settles down upon the deep. Our voyage, like that of the old fabulous navigators of five centuries ago, terminates on the sea in a thick darkness, beyond which there lies no shore and there dawns no light. And it is in the middle of this vast ocean, just where the last zone of the Old Red leans against the first zone of the Silurian, that we have succeeded in discovering a solitary island unseen before,—a shrub-bearing land, much enveloped in fog, but with hills that at least look green in the distance. There are patches of floating sea-weed much comminuted by the surf all around it; and on one projecting headland we see clear through our glasses a cone-bearing tree.

This certainly is not the sort of arrangement demanded by the exigencies of the development hypothesis. A true wood at the base of the Old Red Sandstone, or a true placoid in the Limestones of Bala, very considerably beneath the base of the Lower Silurian system, are untoward misplacements for the purposes of the Lamarckian; and who that has watched the progress of discovery for the last twenty years, and seen the place of the earliest ichthyolite transferred from the Carboniferous to the Cambrian system, and that of the earliest exogenous lignite from the Lias to the Lower Devonian, will now venture to say that fossil wood may not yet be detected as low in the scale as any vegetable organism whatever, or fossil fish as low as the remains of any animal? But though the response of the earlier geologic systems be thus unfavourable to the development hypothesis, may not men such as the author of the "Vestiges" urge, that the geologic evidence, taken as a whole, and in its bearing on groupes and periods, establishes the general fact that the lower plants and animals preceded the higher,—that the conifera, for instance, preceded our true forest trees, such as the oak and elm,—that, in like manner, the fish preceded the reptile, that the reptile preceded the bird, that the bird preceded the mammiferous quadruped and the quadrumana, and that the mammiferous quadruped and the quadrumana preceded man? Assuredly yes! They may and do urge that Geology furnishes evidence of such a succession of existences; and the arrangement seems at once a very wonderful and very beautiful one. Of that great and imposing procession of being of which this world has been the scene, the programme has been admirably marshalled. But the order of the arrangement in no degree justifies the inference based upon it by the Lamarckian. The fact that fishes and reptiles were created on an earlier day than the beasts of the field and the human family, gives no ground whatever for the belief that "the peopling of the earth was one of a natural kind, requiring time," or that the reptiles and fishes have been not only the predecessors, but also the progenitors, of the beasts and of man. The geological phenomena, even had the author of the "Vestiges" been consulted in their arrangement, and permitted to determine their sequence, would yet have failed to furnish, not merely an adequate foundation for the development hypothesis, but even the slightest presumption in its favour. In making good the assertion, may I ask the reader to follow me through the details of a simple though somewhat lengthened illustration?

SUPERPOSITION NOT PARENTAL RELATION.

THE BEGINNINGS OF LIFE.

SEVERAL thousand years ago, ere the upheaval of the last of our raised beaches, there existed somewhere on the British coast a submarine bed, rich in sea-weed and the less destructible zoophytes, and inhabited by the commoner crustaceæ and molluscs. Shoals of herrings frequented it every autumn, haunted by their usual enemies the dog-fish, the cod, and the porpoise; and, during the other seasons of the year, it was swum over by the ling, the hake, and the turbot. A considerable stream, that traversed a wide extent of marshy country, waving with flags and reeds, and in which the frog and the newt bred by millions, entered the sea a few hundred yards away, and bore down, when in flood, its modicum of reptilian remains, some of which, sinking over the submarine bed, found a lodgment at the bottom. Portions of reeds and flags were also occasionally entombed, with now and then boughs of the pine and juniper, swept from the higher grounds. Through frequent depositions of earthy matter brought down by the streamlet, and of sand thrown up by the sea, a gradual elevation of the bottom went on, till at length the deep-sea bed came to exist as a shallow bank, over which birds of the wader family stalked mid-leg deep when plying for food ; and on one occasion a small porpoise, losing his way, and getting entangled amid its shoals, perished on it, and left his carcase to be covered up by its mud and silt. That elevation of the land, or recession of the sea, to which the country owes its last acquired marginal strip of soil, took place, and the shallow bank became a flat meadow, raised some six or eight feet above the sea-level. Herbs, shrubs, and trees, in course of time covered it over ; and then, as century succeeded century, it gathered atop a thick stratum of peaty mould, embedding portions of birch and hazel bushes, and a few doddered oaks. When in this state, at a comparatively recent period, an Italian boy, accompanied by his monkey, was passing over it, when the poor monkey, hard-wrought and ill-fed, and withal but indifferently suited originally for braving the rigours of a keen northern climate, lay down and died, and his sorrowing master covered up the remains. Not many years after, the mutilated corpse of a poor shipwrecked sailor was thrown up, during a night-storm, on the neighbouring beach : it was a mere fragment of the human frame,—a mouldering unsightly mass, decomposing in the sun ; and a humane herd-boy, scooping out a shallow grave for it, immediately over that of the monkey, buried it up. Last of all, a farmer, bent on agricultural improvement, furrowed the flat meadow to the depth of some six or eight feet, by a broad ditch, that laid open its organic contents from top to bottom. And then a philosopher of the school of Maillet and Lamarck, chancing to come the way, stepped aside to examine the phenomena, and square them with his theory.

First, along the bottom of the deep ditch he detects marine organisms of a low order, and generally of a small size.

There are dark indistinct markings traversing the gray silt, which he correctly enough regards as the remains of fucoids; and blent with these he finds the stony cells of flustra, the calcareous spindles of the sea-pen, the spines of echinus, and the thin granular plates of the crustacea. Layers of mussel and pecten shells come next, mixed up with the shells of buccinum, natica, and trochus. Over the shells there occur defensive spines of the dog-fish, blent with the button-like, thornset boucles of the ray. And the minute skeletons of herrings, with the vertebral and cerebral bones of cod, rest over these in turn. He finds also well-preserved bits of reed, and a fragment of pine. Higher up, the well-marked bones of the frog occur, and the minute skeleton of a newt; higher still, the bones of birds of the diver family; higher still, the skeleton of a porpoise; and still higher, he discovers that of a monkey, resting amid the decayed boles and branches of dicotyledonous plants and trees. He pursues his search, vastly delighted to find his doctrine of progressive development so beautifully illustrated; and last of all he detects, only a few inches from the surface, the broken remains of the poor sailor. And, having thus collected his facts, he sets himself to collate them with his hypothesis. To hold that the zoophytes had been created zoophytes, the molluscs molluscs, the fishes fishes, the reptiles reptiles, or the man a man, would be, according to our philosopher, alike derogatory to the Divine wisdom and to the acumen and vigour of the human intellect: it would be "*distressing to him to be compelled to picture the power of God, as put forth in any other manner than in those slow, mysterious, universal laws, which have so plainly an eternity to work in*;" nor, with so large an amount of evidence before him as that which the ditch furnishes,—evidence conclusive to the effect that creation is but development,—does he find it necessary either to cramp his faculties or outrage his taste, by a weak yielding to the requirements of any such belief.

Meanwhile the farmer,—a plain, observant, elderly man, comes up, and he and the philosopher enter into conversation. "I have been reading the history of creation in the side of your deep ditch," says the philosopher, "and find the record really very complete. Look there," he adds, pointing to the unfossiliferous strip that runs along the bottom of the bank; "there, life, both vegetable and animal, first began. It began, struck by electricity out of albumen, as a congeries of minute globe-shaped atoms,—each a hollow sphere within a sphere, as in the well-known Chinese puzzle; and from these living atoms were all the higher forms progressively developed. The ditch, of course, exhibits none of the atoms with which being first commenced; for the atoms don't keep;—we merely see their place indicated by that unfossiliferous band at the bottom; but we may detect immediately over it almost the first organisms into which,—parting thus early into the two great branches of organic being,—they were developed. *There* are the fucoids, first-born among vegetables,—and *there* the zoophytes, well nigh the lowest of the animal forms. The fucoids are marine plants; for, according to Oken, 'all life is from the sea,—none from the continent;' but *there*, a few feet higher, we may see the remains of reeds and flags,—semi-aqueous, semi-aerial plants, of the comparatively low monocotyledonous order into which the fucoids were developed; higher still we detect fragments of pines, and, I think, juniper,—trees and shrubs of the land, of an intermediate order, into which the reeds and flags were developed in turn; and in that peaty layer immediately beneath the vegetable mould, there occur boughs and trunks of blackened oak,—a noble tree of the dicotyledonous division,—the highest to which vegetation in its upward course has yet attained. Nor is the progress of the other great branch of organized being,—that of the animal kingdom,—less distinctly traceable. The zoophytes became crustacea and molluscs,—the crustacea and molluscs, dog-fishes and herrings,—the dog-fish, a low placoid, shot up chiefly into turbot, cod, and ling; but the smaller osseous fish was gradually converted into a batrachian reptile; in short, the herring became a frog,—an animal that still testifies to its ichthyological origin, by commencing life as a fish. Gradually, in the course of years, the reptile, expanding in size and improving in faculty, passed into a warm-blooded porpoise; the porpoise at length, tiring of the water as he began to know better, quitted it altogether, and became a monkey, and the monkey by slow degrees improved into man,—yes, into man, my friend, who has still a tendency, especially when just shooting up to his full stature, and studying the 'Vestiges,' to resume the monkey. Such, Sir, is the true history of creation, as clearly recorded in the section of earth, moss, and silt, which you have so opportunely laid bare. Where that ditch now opens, the generations of the man atop lived, died, and were developed. *There* flourished and decayed his great-great-great-great-grandfather the sea-pen,—his great-great-great-grandfather the mussel,—his great-great-grandfather the herring,—his great-grandfather the frog,—his grandfather the porpoise,—and his father the monkey. And *there* also lived, died, and were developed, the generations of the oak, from the kelp-weed and tangle to the reed and the flag, and from the reed and the flag, to the pine, the juniper, the hazel, and the birch."

"Master," replies the farmer, "I see you are a scholar,

and, I suspect, a wag. It would take a great deal of believing to believe all that. In the days of my poor old neighbour the infidel weaver, who died of *delirium tremens* thirty years ago, I used to read Tom Paine; and, as I was a little wild at the time, I was, I am afraid, a bit of a sceptic. It wasn't easy work always to be as unbelieving as Tom, especially when the conscience within got queasy; but it would be a vast deal easier, Master, to *doubt* with Tom than to *believe* with you. I am a plain man, but not quite a fool; and as I have now been looking about me in this neighbourhood for the last forty years, I have come to know that it gives no assurance that any one thing grew out of any other thing because it chances to be found atop of it, Master. See, yonder is Dobbin lying lazily atop of his bundle of hay; and yonder little Jack, with bridle in hand, and he in a few minutes will be atop of Dobbin. And all I see in that ditch, Master, from top to bottom, is neither more nor less than a certain top-upon-bottom order of things. I see sets of bones and dead plants lying on the top of other sets of bones and dead plants,—things lying atop of things, as I say, like Dobbin on the hay and Jack upon Dobbin. I doubt not the sea was once here, Master, just as it was once where you see the low-lying field yonder, which I won from it ten years ago. I have carted tangle and kelp-weed where I now cut clover and rye-grass, and have gathered periwinkles where I now see snails. But it is *clean against experience*, as my poor old neighbour the weaver used to say,—against *my* experience, Master,—that it was the kelp-weed that became the rye-grass, or that the periwinkles freshened into snails. The kelp-weed and periwinkles belong to those plants and animals of the sea that we find growing in *only* the sea; the rye-grass and snails, to those plants and animals of the land that we find growing on *only* the land. It is contrary to all experience, and all testimony too, that the one passed into the other, and so I cannot believe it; but I do and must believe, instead,—for it is not contrary to experience, and much according to testimony,—that the Author of all created both land productions and sea productions at the 'times before appointed,' and 'determined the bounds of their habitation.' 'By faith we understand that the worlds were framed by the word of God;' and I find I can be a believer on God's terms at a much less expense of credulity than an infidel on yours."

But in this form at least it can be scarce necessary that the argument should be prolonged.

The geological phenomena, I repeat, even had the author of the "Vestiges" been consulted in their arrangement, and permitted to determine their sequence, would fail to furnish a single presumption in favour of the development hypothesis. Does the ditch-side of my illustration furnish it with a single favouring presumption? The arrangement and sequence of the various organisms are complete in both the zoological and phytological branch. The flag and reed succeed the fucoid; the fir and juniper succeed the flag and reed; and the hazel, birch, and oak succeed the fir and juniper. In like manner, and with equal regularity, zoophytes, the radiata, the articulata, mollusca, fishes, reptiles, birds, and mammals, are ranged, the superior in succession over the inferior classes, in the true ascending order; and yet we at once see that the evidence of the ditch-side, amounting in the aggregate to no more than this, that the remains of the higher lie over those of the lower organisms, gives not a shadow of support to the hypothesis that the lower produced the higher. For, according to the honest farmer, the fact that any one thing is found lying on the top of any other thing, furnishes no presumption whatever that the thing below stands in the relation of parent to the thing above. And the evidence which the well-ranged organisms of the ditch-side do not furnish, the organisms of the entire geologic scale, even were they equally well ranged, would fail to supply. The fossiliferous portion of the ditch-side of my illustration may be, let us suppose, some five or six *feet* in thickness; the fossiliferous portion of the earth's crust must be some five or six *miles* in thickness. But the mere circumstance of space introduces no new element into the question. Equally in both cases the fact of superposition is not *identical* with the fact of parental relation, nor even in any degree an *analogous* fact.

As, however, the succession of remains in the fossiliferous series of rocks is infinitely less favourable to the development hypothesis than that of the organisms of the ditch-side, it is not very surprising that the disciples of the development school should be now evincing a disposition to escape from the ascertained facts of Geology, and the legitimate conclusions based upon these, unto unknown and unexplored provinces of the science; or that they should be found virtually urging, that though some of the ascertained facts may seem to bear against them, the facts not yet ascertained may be found telling in their favour. Such, in effect, is the course taken by the author of the "Vestiges," in his "Explanations," when, availing himself of a difference of opinion which exists among some of our most accomplished geologists regarding the first epochs of organized existence, he takes part with the section who hold that we have not yet penetrated to the deposits representative of the dawn of being, and that fossil-charged formations may yet be detected beneath the oldest rocks of what is now regarded as the lowest fossiliferous system. Sir Charles Lyell and Mr Leonard Horner represent the abler and better-known assertors of this last view; while Sir Roderick Murchison and Professor Sedgwick rank among the more distinguished assertors of the antagonist one. It would be of course utterly presumptuous in the writer of these pages to attempt deciding a question regarding which such men differ; but in forming a judgment for myself, various considerations incline me to hold, that the point is now very nearly determined at which, to employ the language of Sir Roderick, "life was first breathed into the waters." The pyramid of organized existence, as it ascends into the by-past eternity, inclines sensibly towards its apex,—that apex of "*beginning*" in which, on far other than geological grounds, it is our privilege to believe. The broad base of the superstructure, planted on the existing *now*, stretches across the entire scale of life, animal and vegetable; but it contracts as it rises into the past;—man,—the quadrumana,—the quadrupedal mammal, —the bird,—and the reptile,—are each in succession struck from off its breadth, till we at length see it with the vertebrata, represented by only the fish, narrowing, as it were, to a point; and though the clouds of the upper region may hide its extreme apex, we infer from the declination of its sides, that it cannot penetrate much farther into the profound. When Steele and Addison were engaged in breaking up, piecemeal, their Spectator Club,—killing off good Sir Roger de Coverly with a defluction, marrying Will Honeycomb to his tenant's daughter, and sending away Captain Sentry and Sir Andrew Freeport to their estates in the country,—it was shrewdly inferred that the "Spectator" himself was very soon to quit the field; and the sudden discontinuance of his lucubrations justified the inference. And a corresponding style of reasoning, based on the corresponding fact of the breaking up and piecemeal disappearance of the group of organized being, seems equally admissible. It is somewhat difficult to conceive how at least *many* more volumes of the geologic record than the known ones could be got up without the *club*. Further,—so far as yet appears, the fish must have lived in advance of the reptile during the three protracted periods of the Old Red Sandstone, the two still more protracted periods of the Upper and Lower Silurians, and the perhaps more protracted period still of the Cambrian deposits;—in all, apparently, a greatly more extended space than that in which the reptile lived in advance of the quadrupedal mammal, or the quadrupedal mammal lived in advance of man. On principles somewhat similar to those on which, with reference to the average term of life, the genealogist fixes the probable period of some birth in his chain of succession of which he cannot determine the exact date, it seems natural to infer that the *birth* of the fish should have taken place at least not earlier than the times of the Cambrian system.

There is another consideration, of at least equal, if not greater weight. A general correspondence is found to obtain in widely-separated localities, in the organic contents of that lowest band of the Lower Silurian or Cambrian system in which fossils have been detected. In Russia, in Sweden, in Norway, in the Lake district of England, and in the United States, there are certain rocks which occupy relatively the same place, and enclose what may be described generally as the same remains. They occur in Scandinavia as that "fucoidal band" of Sir Roderick Murchison which forms the base of the vast Palæozoic basin of the Baltic; they exist in Cumberland and Westmoreland as the Skiddaw slates of Professor Sedgwick, and bear also their fucoidal impressions, blent with graptolites; they are present in North America as those Potsdam sandstones of the States' geologists in which fucoids so abound, mixed with a minute lingula, that they impart to some portions of the strata a carboniferous character. But with these deep-lying beds in all the several localities, thousands of miles apart, in which their passage into the inferior deposits has been traced, fossils cease. And why cease with them? In one locality the ancient ocean may have been of such a depth in the period immediately *previous*, and represented, in consequence, by the strata immediately *beneath*, that no animal could have *lived* at its bottom,—though I do not well see why the remains of those animals who, like the shark and pilot-fish, are frequently seen swimming over the profoundest depths, might not, did such exist at the time, be notwithstanding *found* at its bottom; or in another locality every trace of organization in the nether rocks may have been obliterated, at some posterior period, by fire. But it is difficult to imagine that that uniform cessation of organized life at one point, which seems to have conducted Sir Roderick Murchison and Professor Sedgwick to their conclusion, should have been thus a mere effect of accident. Accident has its laws, but uniformity is not one of them; and should the experience be invariable, as it already seems extensive, that immediately beneath the fucoidal beds organic remains cease, I do not see how the conclusion is to be avoided, that they represent the period in which at least *existences capable of preservation* were first introduced. Every case of coincident cessation which has occurred since the determination of the second case, must be reckoned, not simply as an additional unit in evidence, but, on the principles which determine mathematical probability, as a unit multiplied, first by the chances against its occurrence, regarded as a mere contingency in that exact formation, and second, by the sum of all the previous occurrences at the same point.

In this curious question, however, which it must be the part of future explorers in the geological field definitely to settle, the Lamarckian can have no legitimate stake. It is but natural that, in his anxiety to secure an ultimate retreat for his hypothesis, he should desire to see that darkness in which ghosts love to walk settling down on the extreme verge of the geological horizon, and enveloping in its folds the first beginnings of life. But even did the cloud exist, it is, if I may so express myself, on its nearer side, where there is light, —not within nor beyond it, where there is none,—that the battle must be fought. It is to Geology *as it is known to be* that the Lamarckian has appealed,—not to Geology as it is *not* known to be. He has summoned into court *existing* witnesses; and, finding their testimony unfavourable, he seeks to neutralize their evidence, by calling from the "vasty deep" of the unexamined and the obscure, witnesses that "won't come,"—that by the legitimate authorities are not known even to exist,—and with which he himself is, on his own confession, wholly unacquainted, save in the old scholastic character of mere possibilities. The *possible* fossil can have no more standing in this controversy than the "*possible angel*." He tells us that we have not yet got down to that base-line of all the fossiliferous systems at which life first began ; and very possibly we have not. But what of that? He has carried his appeal to Geology *as it is*;—he has referred his case to the testimony of the *known* witnesses, for in no case can the *unknown* ones be summoned or produced. It is on the evidence of the known, and the known only, that the exact value of his claims must be determined ; and his appeal to the unknown serves but to show how thoroughly he himself feels that the actually ascertained evidence bears against him. The severe censure of Johnson on reasoners of this class is in no degree over-severe. "He who will determine," said the moralist, "against that which he knows, because there may be something which he knows not,—he that can set hypothetical possibility against acknowledged certainty,—is not to be admitted among reasonable beings."

But the honest farmer's reminiscences of his deceased neighbour the weaver, and his use at second-hand of Hume's experience-argument, naturally lead me to another branch of the subject.

LAMARCKIAN HYPOTHESIS OF THE ORIGIN OF PLANTS.

ITS CONSEQUENCES.

I HAVE said that the curiously-mixed, semi-marine, semi-lacustrine flora of the Lake of Stennis became associated in my mind, like the ancient *Asterolepis* of Stromness, with the development hypothesis. The fossil, as has been shown, represents not inadequately the geologic evidence in the question,—the mixed vegetation of the lake may be regarded as forming a portion of the phytological evidence.

"All life," says Oken, "is from the sea. Where the sea organism, by self-elevation, succeeds in attaining into form, there issues forth from it a higher organism. Love arose out of the sea-foam. The primary mucus (that in which electricity originates life) was, and is still, generated in those very parts of the sea where the water is in contact with earth and air, and thus upon the shores. The first creation of the organic took place where the first mountain summits projected out of the water,—indeed, without doubt, in India, if the Himalaya be the highest mountain. *The first organic forms, whether plants or animals, emerged from the shallow parts of the sea.*" Maillet wrote to exactly the same effect a full century ago. "In a word," we find him saying, in his "Telliamed," "do not herbs, plants, roots, grains, and all of this kind that the earth produces and nourishes, come from the sea? Is it not at least natural to think so, since we are certain that all our habitable lands came originally from the sea? Besides, in small islands far from the continent, which have appeared but a few ages ago at most, and where it is manifest that never any man had been, we find shrubs, herbs, roots, and sometimes animals. Now, you must be forced to own either that these productions owed their origin to the sea, or to a new creation, which is absurd."

It is a curious fact, to which, in the passing, I must be permitted to call the attention of the reader, that all the leading assertors of the development hypothesis have been bad geologists. Maillet had for his errors and deficiencies the excellent apology that he wrote more than a hundred years ago, when the theory of a universal ocean, promulgated by Leibnitz nearly a century earlier, was quite as good as any of the other theories of the time, and when Geology, as a science, had no existence. And so we do not wonder at an ignorance which was simply that of his age, when we find him telling his readers that plants *must* have originated in the sea, seeing that "all our habitable lands came originally from the sea;" meaning, of course, by the statement, not at all what the modern geologist would mean were he to employ even the same words, but simply that there was a time when the universal ocean covered the whole globe, and that, as the waters gradually diminished, the loftier mountain summits and higher tablelands, in appearing in their new character as islands and continents, derived their flora from what, in a universal ocean, could be the only possibly existing flora,—that of the sea. But what shall we say of the equally profound ignorance manifested by Professor Oken, a living authority, whom we find prefacing for the Ray Society, in 1847, the English translation of his "Elements of Physio-philosophy?" "The first creation of the organic took place," we find him saying, "where the first mountain summits projected out of the sea,—*indeed, without doubt, in India, if the Himalaya be the highest mountain.*" Here, evidently, in this late age of the world, in which Geology *does* exist as a science, do we find the ghost of the universal ocean of Leibnitz walking once more, as if it had never been laid. Is there now in all Britain even a tyro geologist so unacquainted with geological fact as not to know that the richest flora which the globe ever saw had existed for myriads of ages, and then, becoming extinct, had slept in the fossil state for myriads of ages more, ere the highest summits of the Himalayan range rose over the surface of the deep? The Himalayas disturbed, and bore up along with them in their upheaval, vast beds of the Oolitic system. Belemnites and ammonites have been dug out of their sides along the line of perpetual snow, seventeen thousand feet over the level of the sea. What in the recent period form the loftiest mountains of the globe, existed as portions of a deep-sea bottom, swum over by the fishes and reptiles of the great Secondary period, when what is now Scotland had its dark forests of stately pine,—represented in the present age of the world by the lignites of Helmsdale, Eathie, and Eigg,—and when the plants of a former creation lay dead and buried deep beneath, in shales and fire-clay,—existing as vast beds of coal, or entombed in solid rock, as the brown massy trunks of Granton and Craigleith. And even ere these last existed as living trees, the coniferous lignite of the Lower Old Red Sandstone found at Cromarty had passed into the fossil state, and lay as a semi-calcareous, semi-bituminous mass, amid perished *Dipterians* and extinct *Coccostei.* So much for the Geology of the German Professor. And be it remarked, that the *actualities* in this question can be determined by only the geologist. The mere naturalist may indicate from the analogies of his science, what possibly *might* have taken place; but what really *did* take place, and the true order in which the events occurred, it is the part of the geologist to determine. It cannot be out of place to remark farther, that geological discovery is in no degree responsible for the infidelity of the development hypothesis; seeing that, in the first place, the hypothesis *is greatly more ancient than the discoveries*, and, in the second, that its more prominent assertors are *exactly the men who know least of geological fact.* But to this special point I shall again refer.

The author of the "Vestiges" is at one, regarding the supposed marine origin of terrestrial plants, with Maillet and Oken; and he regards the theory, we find him stating in his "Explanations," as the true key to the well-established fact, that the vegetation of groupes of islands generally corresponds with that of the larger masses of land in their neighbourhood. Marine plants of the same kinds crept out of the sea, it would seem, upon the islands on the one hand, and upon the larger masses of land on the other, and thus produced the same flora in each; just as tadpoles, after passing their transition state, creep out of their canal or river on the opposite banks, and thus give to the fields or meadows on the right-hand side a supply of frogs, of the same appearance and size as those poured out upon the fields and meadows of the left. "Thus, for example," we find him saying, "the Galapagos exhibit general characters in common with South America; and the Cape de Verd islands, with Africa. They are, in Mr Darwin's happy phrase, satellites to those continents, in respect of natural history. Again," he continues, "when masses of land are only divided from each other by narrow seas, there is usually a community of forms. The European and African shores of the Mediterranean present an example. Our own islands afford another of far higher value. It appears that the flora of Ireland and Great Britain is various, or rather that we have five floras or distinct sets of plants, and that each of these is partaken of by a portion of the opposite continent. There are, first, a flora confined to the west of Ireland, and imparted likewise to the north-west of Spain; second, a flora in the south-west promontory of England and of Ireland, extending across the Channel to the north-west coast of France; third, one common to the south-east of England and north of France; fourth, an Alpine flora developed in the Scottish and Welsh Highlands, and intimately related to that of the Norwegian Alps; fifth, a flora which prevails over a large part of England and Ireland, 'mingled with other floras, and diminishing slightly as we proceed westward:' this bears intimate relation with the flora of Germany. Facts so remarkable would force the meanest fact-collector or species-demonstrator into generalization. The really ingenious man who lately brought them under notice (Professor Edward Forbes) could only surmise, as their explanation, that the spaces now occupied by the intermediate seas must have been dry land at the time when these floras were created. In that case, either the original arrangement of the floras, or the selection of land for submergence, must have been apposite to the case in a degree far from usual The necessity for a simpler cause is obvious, and it is found in the hypothesis of *a spread of terrestrial vegetation from the sea into the lands adjacent.* The community of forms in the various regions opposed to each other merely indicates a distinct marine creation in each of the oceanic areas respectively interposed, and which would naturally advance into the lands nearest to it, as far as circumstances of soil and climate were found agreeable."

Such, regarding the origin of terrestrial vegetation, are the views of Maillet, Oken, and the author of the "Vestiges." They all agree in holding that the plants of the land existed in their first condition as weeds of the sea.

Let me request the reader, at this stage, ere we pass on to the consideration of the experience-argument, to remark a few incidental, but by no means unimportant, consequences of the belief. And, first, let him weigh for a moment the comparative demands on his credulity of the theory by which Professor Forbes accounts for the various floras of the British Islands, and that hypothesis of transmutation which the author of the "Vestiges" would so fain put in its place, as greatly more simple, and, of course, more in accordance with the principles of human belief. In order to the reception of the Professor's theory, it is necessary to hold, in the first place, that the creation of each species of plant took place, not by repetition of production in various widely-separated centres, but in some single centre, from which the species propagated itself by seed, bud, or scion, across the special area which it is now found to occupy. And this, in the first instance, is of course as much an assumption as any of those assumed numbers or assumed lines with which, in algebra and the mathematics, it is necessary in so many calculations to set out, in quest of some required number or line, which, without the assistance of the assumed ones, we might despair of ever finding. But the assumption is in itself neither unnatural nor violent; there are various very remarkable analogies which lend it support ; the facts which seem least to harmonize with it are not wholly irreconcileable, and are, besides, of a merely exceptional character; and, further, it has been adopted by botanists of the highest standing.* It

---

* The following digest, from Professor Balfour's very admirable "Manual of Botany," of what is held on this curious subject, may be not unacceptable to the reader. "It is an interesting question to determine the mode in which the various species and tribes of plants were originally scattered over the globe. Various hypotheses have been advanced on the subject. Linnæus entertained the opinion that there was at first only one primitive centre of vegetation, from which plants were distributed over the globe. Some, avoiding all discussions and difficulties, suppose that plants were produced at first in the localities where they are now seen vegetating. Others think that each species of plant originated in, and was diffused from, a single primitive centre; and that there were numerous such centres situated in different parts of the world, each centre being the seat of a particular number of species. They thus admit great vegetable migrations, similar to those of the human races. Those who adopt the latter view recognise in the distribution of plants some of the last revolutions of our planet, and the action of numerous and varied forces, which impede or favour the dissemination of vegetables in the present day. They endeavour to ascertain the primitive flora of countries, and to trace the vegetable migrations which have taken place. Daubeny says, that analogy favours the supposition that each species of plant was originally formed in some particular locality, whence it spread itself gradually over a certain area, rather than that the earth was at once, by the fiat of the Almighty, covered with vegetation in the manner we at present behold it. The human race rose from a single pair ; and the distribution of plants and animals over a certain definite area would seem to imply that the same was the general law. Analogy would lead us to believe that the extension of species over the earth originally took place on the same plan on which it is conducted at present, when a new island starts up in the midst of the ocean, produced either by a coral reef or a volcano. In these cases the whole surface is not at once overspread with plants, but

is necessary to hold, in the second place, in order to the reception of the theory, that the area of the earth's surface occupied by the British islands and the neighbouring coasts of the Continent once stood fifty fathoms higher, in relation to the existing sea-level, than it does now,—a belief which, whatever its specific grounds or standing in this particular case, is at least in strict accordance with the general geological phenomena of subsidence and elevation, and which, so far from outraging any experience founded on observation or testimony, runs in the same track with what is known of wide areas now in the course of sinking, like that on the Italian coast, in which the Bay of Baiæ and the ruins of the temple of Serapis occur, or that in Asia, which includes the Run of Cutch; or of what is known of areas in the course of rising, like part of the coast of Sweden, or part of the coast of South America, or in Asia along the western shores of Aracan. Whereas, in order to close with the *simpler* antagonistic belief of the author of the "Vestiges," it is necessary to hold, *contrary* to all experience, that *dulce* and *hen-ware** became, through a very wonderful metamorphosis, cabbage and spinnage; that kelp-weed and tangle bourgeoned into oaks and willows; and that *slack*, *rope-weed*, and *green-raw*,† shot up into mangel-wurzel, rye-grass, and clover. *Simple*, certainly! An infidel on terms such as these could with no propriety be regarded as an *unbeliever*. It is well

---

a gradual progress of vegetation is traced from the accidental introduction of a single seed, perhaps, of each species, wafted by winds or floated by currents. The remarkable limitation of certain species to single spots on the globe seems to favour the supposition of specific centres."

* *Rhodymenia palmata* and *Alaria esculenta*.

† *Porphyra laciniata*, *Chorda filum*, and *Enteromorpha compressa*.

that the New Testament makes no such extraordinary demands on human credulity.

Let us remark further, at this stage, that, judging from the generally received geological evidence in the case, very little time seems to be allowed by the author of the "Vestiges" for that miraculous process of transmutation through which the low algæ of our sea-shores are held to have passed into the high orders of plants which constitute the prevailing British flora. The boulder clay, which rises so high along our hills, and which, as shown by its inferior position on the lower grounds, is decidedly the most ancient of the country's superficial deposits, is yet so modern, geologically, that it contains only recent shells. It belongs to that cold, glacial, post-Tertiary period, in which what is now Britain existed as a few groupes of insulated hill-tops, bearing the semi-arctic vegetation of our fourth flora,—that true *Celtic* flora of the country which we now find, like the country's Celtic races of our own species, cooped up among the mountains. The fifth or Germanic flora must have been introduced, it is held, at a later period, when the climate had greatly meliorated. And if we are to hold that the plants of this last flora were *developed* from sea-weed, not propagated across a continuity of land from the original centre in Germany, or borne by currents from the mouths of the Germanic rivers,—the theory of Mon. C. Martins,—then must we also hold that that development took place since the times of the boulder clay, and that fucoids and confervæ became dicotyledonous and monocotyledonous plants during a brief period, in which the *Purpura lapillus* and *Turritella terebra* did not alter a single whorl, and the *Cyprina Islandica* and *Astarte Borealis* retained unchanged each minute projection of their hinges, and each nicer peculiarity of their muscular impressions. *Creation* would be greatly less wonderful than a sudden transmutative process such as this, restricted in its operation to groupes of English, Irish, and Manx plants, identical with groupes in Germany, when all the various organisms around them, such as our sea-shells, continued to be exactly what they had been for ages before. A process of development from the lowest to the highest forms, rigidly restricted to the flora of a country, would be simply the miracle of Jonah's gourd several thousand times repeated.

I must here indulge in a few remarks more, which, though they may seem of an incidental character, have a direct bearing on the general subject. The geologist infers, in all his reasonings founded on fossils, that a race or species has existed from some one certain point in the scale to some other certain point, if he find it occurring at both points together. He infers on this principle, for instance, that the boulder clay, which contains only *recent* shells, belongs to the *recent* or post-Tertiary period ; and that the Oolite and Lias, which contain *no* recent shells, represent a period whose existences have all become extinct. And all experience serves to show that his principle is a sound one. In creation there are many species linked together, from their degree of similarity, by the *generic* tie ; but no perfect verisimilitude obtains among them, unless hereditarily derived from the one, two, or more individuals, of contemporary origin, with which the race began. True, there are some races that have spread over very wide circles,—the circle of the human family has become identical with that of the globe ; and there are certain plants and animals that, from peculiar powers of adaptation to the varieties of soil and climate,—mayhap also from the tenacious vitality of their seeds, and their facilities of transport by natural means,—are likewise diffused very widely. There are plants, too, such as the common nettle and some of the ordinary grasses, which accompany civilized man all over the globe, he scarce knows how, and spring up unbidden whereever he fixes his habitation. He, besides, carries with him the common agricultural weeds : there are localities in the United States, says Sir Charles Lyell, where these *exotics* outnumber the native plants; but these are exceptions to the prevailing economy of distribution; and the circles of species generally are comparatively limited and well defined. The mountains of the southern hemisphere have, like those of Switzerland and the Scotch Highlands, their forests of coniferous trees; but they furnish no Swiss pines or Scotch firs; nor do the coasts of New Zealand or Van Dieman's Land supply the European shells or fish. True, there may be much to puzzle in the identity of what may be termed the exceptional plants, equally indigenous, apparently, in circles widely separated by space. It has been estimated that there exist about a hundred thousand vegetable species, and of these, thirty Antarctic forms have been recognised by Dr Hooker as identical with European ones. Had Robinson Crusoe failed to remember that he had shaken the old corn-bag where he found the wheat and barley ears springing up on his island, he might have held that he had discovered a new centre of the European ceralia. And the process analogous to the shaking of the bag is frequently a process *not* to be remembered. There are several minute lochans in the Hebrides and the west of Ireland in which there occurs a small plant of the cord-rush family (*Eriocaulon Septangulare*), which, though common in America, is nowhere to be found on the European Continent. It is the only British plant which belongs to no other part of Europe. How was it transported across the Atlantic? Entangled, mayhap, in the form of a single seed,—for its seeds are exceedingly light and small, —in the plumage of some water-fowl, free of both sea and lake, it had been carried in the germ from the weed-skirted edge of some American swamp or mere, to some mossy lochan of Connaught or of Skye; and one such seed transported by one such accident, unique in its occurrence in thousands of years, would be quite sufficient to puzzle all the botanists for ever after. I have seen the seed of one of our Scotch grasses, that had been originally caught in the matted fleece of a sheep reared among the hills of Sutherland, and then wrought into a coarse, ill-dressed woollen cloth, carried about for months in a piece of underclothing. It might have gone over half the globe in that time, and, when cast away with the worn vestment, might have originated a new circle for its species in South America or New Holland. There are seeds specially contrived by the Great Designer to be carried far from their original habitats in the coats of animals,—a mode which admits of transport to much greater distances than the mode, also extensively operative, of consigning them for conveyance to their stomachs; and when we see the work in its effects, we are puzzled by the want of a record of an emigratory process, of which, in the circumstances, no record could possibly exist. Unable to make out a case for the "shaking of the bag," we bethink us, in the emergency, of repetition of creation. But in circles separated by *time*, not space,—by *time*, across whose dim gulfs no voyager sails and no bird flies, and over which there are no means of transport from the point where a race once fails, to any other point in the future,—we find no repetition of species. If the production of perfect duplicates or triplicates in independent centres were a law of nature, our works of physical science could scarce fail to tell us of identical species found occurring in widely-separated systems,—Scotch firs and larches, for instance, among the lignites of the Lias, or *Cyprina Islandica* and *Ostrea edulis* among the shells of the Mountain Limestone. But never yet has the geologist found in his systems or formations any such evidence as facts such as these might be legitimately held to furnish, of the independent *de novo* production of individual members of any single species. On the contrary, the evidence lies so entirely the other way, that he reasons on the existence of a family relation obtaining between all the members of each species, as one of his best established principles. If members of the same species may exist through *de novo* production, without hereditary relationship, so thoroughly, in consequence, does the fabric of geological reasoning fall to the ground, that we find ourselves incapacitated from regarding even the bed of common cockle or mussel shells, which we find lying a few feet from the surface on our raised beaches, as of the existing creation at all. Nay, even the human remains of our moors may have belonged, if our principle of relationship in each species be not a true one, to some former creation, cut off from that to which we ourselves belong, by a wide period of death. All palæontological reasoning is at an end for ever, if identical species can originate in independent centres, widely separated from each other by periods of time; and if they fail to originate in periods separated by time, how or why in centres separated by space?

Let the reader remark further, the bearing of those facts from which this principle of geological reasoning has been derived, on the development hypothesis. We find species restricted to circles and periods; and though stragglers are occasionally found outside their circle in the existing state of things, never are they found beyond their period among the remains of the past. It was profoundly argued by Cuvier, that *life* could not possibly have had a chemical origin. "In fact," we find him remarking, "life exercising upon the elements which at every instant form part of the living body, and upon those which it attracts to it, an action contrary to that which would be produced without it by the usual chemical affinities, it is inconsistent to suppose that it can itself be produced by these affinities." And the phenomena of restriction to circle and period testify to the same effect. Nothing, on the one hand, can be more various in character and aspect than the organized existences of the various circles and periods ; nothing more invariable, on the other, than the results of chemical or electrical experiment. And yet, to use almost the words of Cuvier, "we know of no other power in nature capable of re-uniting previously separated molecules," than the electric and the chemical. To these agents, accordingly, all the assertors of the development hypothesis have had recourse for at least the *origination* of life. Air, water, earth existing as a saline mucus, and an active persistent electricity, are the creative ingredients of Oken. The author of the "Vestiges" is rather less explicit on the subject : he simply refers to the fact, that the "basis of all vegetable and animal substances consists of nucleated cells,—that is, of cells having granules within them ;" and states that globules of a resembling character "can be produced in albumen by electricity ;" and that though albumen itself has not yet been produced by artificial means,—the only step in the process of creation which is wanting,—it is yet known to be a chemical composition, the mode of whose production may "be any day discovered in the laboratory." Further, he adopts, as part of the foundation of his hypothesis, the pseudo-experiment of Mr Weekes, who holds that out of certain saline preparations, acted upon by electricity, he can produce certain living animalcula of the mite family;—the vital and the organized out of the inorganic and the dead. In all such cases, electricity, or rather, according to Oken, galvanism, is regarded as the vitalizing principle. "*Organism*," says the German, "is *galvanism* residing in a thoroughly homogeneous mass. . . . . A galvanic pile pounded into atoms must become alive. In this manner nature brings forth organic bodies." I have even heard it seriously asked whether electricity be not God! Alas! could such a god, limited in its capacity of action, like those "gods of the plains" in which the old Syrian trusted, have wrought, in the character of Creator, with a variety of result so endless, that in no geologic period has repetition taken place? In all that purports to be experiment on the development side of the question, we see nothing else save repetition. The *Acarus Crossii* of Mr Weekes is not a new species, but the *repetition* of an old one, which has been long known as the *Acarus horridus*, a little bristle-covered creature of the mite family, that harbours in damp corners among the debris of outhouses, and the dust and dirt of neglected work-shops and laboratories. Nay, even a change in the chemical portion of the experiment by which he believed the creature to be produced, failed to secure variety. A powerful electric current had been sent, in the first instance, through a solution of silicate of potash, and, after a time, the *Acarus horridus* crawled out of the fluid. The current was then sent through a solution of nitrate of copper, and, after a due space, the *Acarus horridus* again creeped out. A solution of ferro-cyanate of potash was next subjected to the current, and yet again, and in greater numbers than on the two former occasions, there appeared, as in virtue, it would seem, of its extraordinary appetency, *to be* the same ever-recurring *Acarus horridus.* How, or in what form, the little creature should have been introduced into the several experiments, it is not the part of those who question their legitimacy to explain: it is enough for us to know, that individuals of the family to which the *Acarus* belongs are so remarkable for their powers of life, even in their fully developed state, as to resist, for a time, the application of boiling water, and to live long in alcohol. We know, further, that the *germs* of the lower animals are greatly more tenacious of vitality than the animals themselves; and that they may exist in their state of embryoism in the most unthought of and elusive forms; nay,—as the recent discoveries regarding alterations of generation have conclusively shown,—that the germ which produced the parent may be wholly unlike the germ that produces its offspring, and yet identical with that which produced the parent's parent. Save on the theory of a quiescent vitality, maintained by seeds for centuries within a few inches of the earth's surface, we know not how a layer of shell, sand, or marl, spread over the bleak moors of Harris, should produce crops of white clover, where only heath had grown before; nor how brakes of doddered furze burnt down on the slopes of the Cromarty Sutors should be so frequently succeeded by thickets of raspberry. We are not, however, to give up the *unknown,*—that illimitable province in which science discovers,—to be a wild region of dream, in which fantasy may invent. There are many dark places in the field of human knowledge which even the researches of ages may fail wholly to enlighten; but no one derives a right from that circumstance to people them with chimeras and phantoms. They belong to the philosophers of the future,—not to the visionaries of the present. But while it is not our part to explain *how*, in the experiments of Mr Weekes, the chain of life from life has been maintained unbroken, we can most conclusively show, that that world of organized existence of which we ourselves form part, is, and ever has been, a world, not of tame repetition, but of endless variety. It is palpably not a world of *Acaridæ* of one species, nor yet of creatures developed from these, under those electric or chemical laws of which the grand characteristic is invariability of result. The vast variety of its existences speak not of the operation of *unvarying laws*, that represent, in their uniformity of result, the unchangeableness of the Divinity, but of *creative acts*, that exemplify the infinity of His resources.

Let the reader yet farther remark, if he has followed me through these preliminary observations, what is really involved in the hypothesis of the author of the "Vestiges," regarding the various floras common to the British islands and the Continent. If it was upon his scheme that England, Ireland, and the mainland of Europe came to possess an identical flora, production *de novo* and by repetition of the same species must have taken place in thousands of instances along the shores of each island and of the mainland. His hypothesis demands that the sea-weed on the coast of Ireland should have been developed, first through lower, and then higher forms, into thousands of terrestrial plants,—that exactly the same process of development from sea-weed into terrestrial plants of the same species should have taken place on the coast of England, and again on the coasts of the Continent generally,—and that identically the same vegetation should have been originated in this way in at least three great centres. And if plants of the same species could have had three distinct centres of organization and development, why not three hundred, or three thousand, or three hundred thousand? Nor will it do to attempt escaping from the difficulty, by alleging that there is the groundwork in the case of at least a common marine vegetation to start from; and that thus, if we have not properly the existence of the direct hereditary tie among the various individuals of each species, we may yet recognise at least a sort of collateral relationship among them, derived from the relationship of their marine ancestry. For relationship, in even the primary stage, the author of the "Vestiges" virtually repudiates, by adopting, as one of the foundations of his hypothesis, with, of course, all the legitimate consequences, the experiments of Mr Weekes. The animalculæ-making process is instanced as representative of the first stage of being,—that in which dead inorganic matter assumes vitality; and it corresponds, in the zoological branch, to the production of a low marine vegetation in the phytological one. A certain semi-chemical, semi-electrical process, originates, time after time, certain numerous low forms of life, identical in species, but connected by no tie of relationship: such is the presumed result of the Weekes experiment. A certain farther process of development matures low forms of life, thus originated, into higher species, also identical, and also wholly unconnected by the family tie: such are the consequences legitimately involved in that island-vegetation theory promulgated by the author of the "Vestiges." And be it remembered, that Mr Weekes' process, so far as it is simply electrical and chemical, is a process which is as capable of having been gone through in all times and all places, as that other process of strewing marl upon a moor, through which certain rustic experimenters have held that they produced white clover. It could have been gone through during the Carboniferous or the Silurian period; for all truly chemical and electrical experiments would have resulted in manifestations of the same phenomena then as now ;—an acid would have effervesced as freely with an alkali ; and each fibre of an electrified feather,—had feathers then existed,—would have stood out as decidedly apart from all its neighbours. We must therefore hold, if we believe with the author of the "Vestiges," first, from the Weekes experiment, that in all times, and in all places, every centre of a certain chemical and electric action would have become a new centre of creation to certain *recent* species of low, but not *very* low, organization ; and, second, from his doctrine regarding the identity of the British and Continental floras, that in the course of subsequent development from these low forms, the process in each of many widely-separated centres,—widely separated both by space and time,—would be so nicely correspondent with the process in all the others, that the same higher *recent* forms would be matured in all. And to doctrines such as these, the experience of all Geologists, all Phytologists, all Zoologists, is diametrically opposed. If these doctrines be true, *their* sciences are false in their facts, and idle and unfounded in their principles.

THE TWO FLORAS, MARINE AND TERRESTRIAL.

BEARING OF THE EXPERIENCE ARGUMENT.

Is the reader acquainted with the graphic verse, and scarce less graphic prose, in which Crabbe describes the appearances presented by a terrestrial vegetation affected by the waters of the sea? In both passages, as in all his purely descriptive writings, there is a solidity of truthful observation exhibited, which triumphs over their general homeliness of vein.

> "On either side
> Is level fen, a prospect wild and wide,
> With dykes on either hand, by ocean self-supplied.
> Far on the right the distant sea is seen,
> And salt the springs that feed the marsh between;
> Beneath an ancient bridge the straitened flood
> Rolls through its sloping banks of slimy mud;
> Near it a sunken boat resists the tide,
> That frets and hurries to the opposing side;
> The rushes sharp, that on the borders grow,
> Bend their brown florets to the stream below,
> Impure in all its course, in all its progress slow.
> Here a grave Flora scarcely deigns to bloom,
> Nor wears a rosy blush, nor sheds perfume.
> The few dull flowers that o'er the place are spread,
> Partake the nature of their fenny bed;
> Here on its wiry stem, in rigid bloom,
> Grows the salt lavender, that lacks perfume;

> Here the dwarf sallows creep, the septfoil harsh,
> And the soft slimy mallow of the marsh.
> Low on the ear the distant billows sound,
> And just in view appears their stony bound."

"The ditches of a fen so near the ocean," says the poet, in the note which accompanies this passage, "are lined with irregular patches of a coarse-stained laver ; a muddy sediment rests on the horse-tail and other perennial herbs which in part conceal the shallowness of the stream ; a fat-leaved, pale-flowering scurvy-grass appears early in the year, and the razor-edged bullrush in the summer and autumn. The fen itself has a dark and saline herbage : there are rushes and *arrow-head ;* and in a few patches the flakes of the cotton-grass are seen, but more commonly the *sea-aster*, the dullest of that numerous and hardy genus ; a *thrift*, blue in flower, but withering, and remaining withered till the winter scatters it ; the *salt-wort*, both simple and shrubby ; a few kinds of grass changed by the soil and atmosphere ; and low plants of two or three denominations, undistinguished in the general view of scenery ;—such is the vegetation of the fen where it is at a small distance from the ocean."

And such are the descriptions of Crabbe, at once a poet and a botanist. In referring to the blue tint exhibited in salt fens by the pink-coloured flower of the *thrift* (*Statice Armeria*), he might have added, that the general green of the terrestrial vegetation likewise assumes, when subjected to those modified marine influences under which plants of the land can continue to live, a decided tinge of blue. It is further noticeable, that the general brown of at least the larger algæ presents, as they creep upwards upon the beach to meet with these, a marked tinge of yellow. The prevailing brown of the one flora approximates towards yellow,—the prevailing green of the other towards blue; and thus, instead of mutually merging into some neutral tint, they assume at their line of meeting directly antagonistic hues.

But what does experience say regarding the transmutative conversion of a marine into a terrestrial vegetation,—that experience on which the sceptic founds so much? As I walked along the green edge of the Lake of Stennis, selvaged by the line of detached weeds with which a recent gale had strewed its shores, and marked that for the first few miles the accumulation consisted of marine algæ, here and there mixed with tufts of stunted reeds or rushes, and that as I receded from the sea it was the algæ that became stunted and dwarfish, and that the reeds, aquatic grasses, and rushes, grown greatly more bulky in the mass, were also more fully developed individually, till at length the marine vegetation altogether disappeared, and the vegetable debris of the shore became purely lacustrine,—I asked myself whether here, if anywhere, a transition flora between lake and sea ought not to be found? For many thousand years ere the tall gray obelisks of Stennis, whose forms I saw this morning reflected in the water, had been torn from the quarry, or laid down in mystic circle on their flat promontories, had this lake admitted the waters of the sea, and been salt in its lower reaches and fresh in its higher. And during this protracted period had its quiet, well-sheltered bottom been exposed to no disturbing influences through which the delicate process of transmutation could have been marred or arrested. Here, then, if in any circumstances, ought we to have had, in the broad permanently brackish reaches, at least indications of a vegetation intermediate in its nature between the monocotyledons of the lake and the algæ of the sea; and yet not a vestige of such an intermediate vegetation could I find among the up-piled debris of the mixed floras, marine and lacustrine. The lake possesses no such intermediate vegetation. As the water freshens in its middle reaches, the algæ become dwarfish and ill-developed ; one species after another ceases to appear, as the habitat becomes wholly unfavourable to it ; until at length we find, instead of the brown, rootless, flowerless fucoids and confervæ of the ocean, the green, rooted, flower-bearing flags, rushes, and aquatic grasses of the fresh water. Many thousands of years have failed to originate a single intermediate plant. And such, tested by a singularly extensive experience, is the general evidence.

There is scarce a chain-length of the shores of Britain and Ireland that has not been a hundred and a hundred times explored by the botanist,—keen to collect and prompt to register every rarity of the vegetable kingdom ; but has he ever yet succeeded in transferring to his herbarium a single plant caught in the transition state ? Nay, are there any of the laws under which the vegetable kingdom exists better known than those laws which fix certain species of the algæ to certain zones of coast, in which each, according to the overlying depth of water and the nature of the bottom, finds the only habitat in which it can exist ? The rough-stemmed tangle (*Laminaria digitata*) can exist no higher on the shore than the low line of ebb during stream-tides ; the smooth-stemmed tangle (*Laminaria saccharina*) flourishes along an inner belt, partially uncovered during the ebbs of the larger neaps ; the forked and cracker kelp-weeds (*Fucus serratus* and *Fucus nodosus*) thrive in a zone still less deeply covered by water, and which even the lower neaps expose. And at least one other species of kelp-weed, the *Fucus vesiculosus*, occurs in a zone higher still, though, as it creeps upwards on the rocky beach, it loses its characteristic bladders, and becomes short and narrow of frond. The thick brown tufts of *Fucus canaliculatus*, which in the lower and middle reaches of the Lake of Stennis I found heaped up in great abundance along the shores, also rises high on rocky beaches,—so high in some instances, that during neap-tides it remains uncovered by the water for days together. If, as is not uncommon, there be an escape of land-springs along the beach, there may be found, where the fresh water oozes out through the sand and gravel, an upper terminal zone of the confervæ, chiefly of a green colour, mixed with the ribbon-like green laver (*Ulva latissima*), the purplish-brown laver (*Porphyra laciniata*), and still more largely with the green silky Enteromorpha (*E. compressa*).* And then, decidedly within the line of the storm-beaches of winter,—not unfrequently in low sheltered bays, such as the Bay of Udale or of Nigg, where the ripple of every higher flood washes,—we may find the vegetation of the land,—represented by the sentinels and picquets of its outposts,—coming down, as if to meet with the higher-growing plants of the sea. In salt marshes the two vegetations may be seen, if I may so express myself, *dovetailed* together at their edges,—at least one species of club-rush (*Scirpus maritimus*) and the common saltwort and glasswort (*Salsola kali* and *Salicornia procumbens*) encroaching so far upon the sea as to mingle with a thinly-

* "Dr Neill mentions," says the Rev. Mr Landsborough, in his complete and very interesting "History of British Sea-Weeds," "that on our shores algæ generally occupy zones in the following order, beginning from deep water:—*F. filum*; *F. esculentus* and *bulbosus*; *F. digitatus*, *saccharinus*, and *loreus*; *F. serratus* and *crispus*; *F. nodosus* and *vesiculosus*; *F. canaliculatus*; and, last of all, *F. pygmæus*, which is satisfied if it be within reach of the spray."

scattered and sorely-diminished fucus,—that bladderless variety of the *Fucus vesiculosus* to which I have already referred, and which may be detected in such localities, shooting forth its minute brown fronds from the pebbles. On rocky coasts, where springs of fresh water come trickling down along the fissures of the precipices, the observer may see a variety of *Rhodomenia palmata*,—the fresh-water dulse of the Moray Frith,—creeping upwards from the lower limits of production, till just where the common gray balanus ceases to grow. And there, short and thick, and of a bleached yellow hue, *it* ceases also ; but one of the commoner marine confervæ,—the *Conferva arcta*, blent with a dwarfed *Enteromorpha*,—commencing a very little below where the dulse ends, and taking its place, clothes over the runnels with its covering of green for several feet higher : in some cases, where it is frequently washed by the upward dash of the waves, it rises above even the flood-line ; and in some crevice of the rock beside it, often as low as its upper edge, we may detect stunted tufts of the sea-pink or of the scurvy-grass. But while there is thus a vegetation intermediate *in place* between the land and the sea, we find, as if it had been selected purposely to confound the transmutation theory, that it is in no degree intermediate in character. For, while it is chiefly marine weeds of the lower division of the confervæ that creep upwards from the sea to meet the vegetation of the land, it is chiefly terrestrial plants of the higher division of the dicotyledons that creep downwards from the land to meet the vegetation of the sea. The salt-worts, the glass-worts, the arenaria, the thrift, and the scurvy grass, are all dicotyledonous plants. Nature draws a deeply-marked line of division where the requirements of the transmutative hypothesis would demand the nicely graduated softness of a shaded one; and, addressing the strongly marked floras on either hand, even more sternly than the waves themselves, demands that to a certain definite bourne should they come, and no farther.

But in what form, it may be asked, or with what limitations, ought the Christian controversialist to avail himself, in this question, of the experience argument? Much ought to depend, I reply, on the position taken up by the opposite side. We find no direct reference made by the author of the "Vestiges" to the anti-miracle argument, first broached by Hume, in a purely metaphysical shape, in his well-known "Inquiry," and afterwards thrown into the algebraic form by La Place, in his *Essai Philosophique sur les Probabilités.* But we do detect its influences operative throughout the entire work. It is because of some felt impracticability on the part of its author, of attaining to the prevailing belief in the *miracle* of creation, that he has recourse, instead, to the so-called *law* of development. The *law* and the *miracle* are the alternatives placed before him; and, rejecting the *miracle*, he closes with the *law*. Now, in such circumstances, he can have no more cause of complaint, if, presenting him with the experience argument of Hume and La Place, we demand that he square the evidence regarding the existence of his *law* strictly according to its requirements, than the soldier of an army that charged its field-pieces with rusty nails would have cause of complaint if he found himself wounded by a missile of a similar kind, sent against him by the artillery of the enemy. You cannot, it might be fairly said, in addressing him, acquiesce in the miracle here, because, as a violation of the laws of nature, there are certain objections, founded on invariable experience, which bear direct against your belief in it. Well, here are the objections, in the strongest form in which they have yet been stated; and here is your hypothesis respecting the development of marine algæ into terrestrial plants. We hold that against that hypothesis the objections bear at least as directly as against any miracle whatever,—nay, that not only is it contrary to an invariable experience, but opposed also to all testimony. We regard it as a mere idle dream. Maillet dreamed it,—and Lamarck dreamed it,—and Oken dreamed it; but none of them did more than merely dream it: its existence rests on exactly the same basis of evidence as that of Whang the miller's "monstrous pot of gold and diamonds," of which he dreamed three nights in succession, but which he never succeeded in finding. If we are in error in our estimate, here is the argument, and here the hypothesis: give us, in support of the hypothesis, the amount of evidence, founded on a solid experience, which the argument demands.

But to leave the experience argument in exactly the state in which it was left by Hume and La Place, would be doing no real justice to our subject. It is in that state quite sufficient to establish the fact, that there can be no real escape from belief in *acts of creation* never witnessed by man, to *processes of development* never witnessed by man; seeing that a presumed *law* beyond the cognizance of experience must be as certainly rejected, on the principle of the argument, as a presumed *miracle* beyond that cognizance. It places the presumed *law* and the presumed *miracle* on exactly the same level. But there is a palpable flaw in the anti-miracle argument. It does not prove that miracles *may not have taken place*, but that miracles, whether they have taken place or no, are *not to be credited*, and this simply because they *are* miracles, *i.e.* violations of the established laws of nature. And if it be possible for events to take place which man, on certain principles, is imperatively required not to credit, these principles must of course serve merely to establish a discrepancy between the actual *state* of things, and what is to be *believed* regarding it. And thus, instead of serving purposes of truth, they are made to subserve purposes of error; for the existence of truth in the mind is neither more nor less than the existence of certain conceptions and beliefs, adequately representative of what actually *is*, or what really *has taken place*.

I cannot better illustrate this direct tendency of the anti-miracle argument to destroy truth in the mind, by bringing the mental beliefs into a state of non conformity with the possible and actual, than by a quotation from La Place himself:—"We would not," he says, "give credit to a man who would affirm that he saw a hundred dice thrown into the air, and that they all fell on the same faces. If we had ourselves been spectators of such an event, we would not believe our own eyes till we had scrupulously examined all the circumstances, and assured ourselves that there was no trick or deception. After such an examination, we would not hesitate to admit it, notwithstanding its great improbability: and no one would have recourse to an inversion of the laws of vision in order to account for it." Now, here is the principle broadly laid down, that it is impossible to communicate by the evidence of testimony, belief in an event which *might* happen, and which, if it happened, *ought* on certain conditions to be credited. No one knew better than La Place himself, that the *possibility* of the event which he instanced could be represented with the utmost exactitude by figures. The probability, in throwing a single die, that the ace will be presented on its upper face, is as one in six,—six being the entire number of sides which the cube can possibly present, and the side with the ace being one of these;—the probability that in throwing a *pair* of dice the aces of both will be at once presented on their upper faces, is as one in thirty-six, as against the one-sixth chance of the ace being presented by the one, there are also six chances that the ace of the other should not concur with it ;—and in throwing *three* dice, the probability that their three aces should be at once presented is, of course, on the same principle, as one in six times thirty-six, or, in other words, as one in two hundred and sixteen. And thus, in ascertaining the exact degree of probability of the hundred aces at once turning up, we have to go on multiplying by six, for each die we add to the number, the product of the immediately-previous calculation. Unquestionably, the number of chances *against*, thus balanced with the single chance *for*, would be very great ; but its existence as a definite number would establish, with all the force of arithmetical demonstration, the *possibility* of the event ; and if an eternity were to be devoted to the throwing into the air of the hundred dice, it would occur an *infinite number of times.* And yet the principle of Hume and La Place forms, when adopted, an impassable gulf between this possibility and human belief. The possibility might be embodied, as we see, in an actual occurrence,—an occurrence witnessed by hundreds ; and yet the anti-miracle argument, as illustrated by La Place, would cut off all communication regarding it between these hundreds of witnesses, however unexceptionable their character as such, and the rest of mankind. The principle, instead of giving us a right rule through which the beliefs in the mind are to be rendered correspondent with the reality of things, goes merely to establish a certain imperfection of transmission from one mind to another, in consequence of which, realities in fact, if very extraordinary ones, could not possibly be received as objects of belief, nor the mental appreciation of things be rendered adequately concurrent with the state in which the things really existed.

Nor is the case different when, for a *possibility* which the arithmetician can represent by figures, we substitute the *miracle* proper. Neither Hume nor La Place ever attempted to show that miracles could not take place; they merely directed their argument against a belief in them. The wildest sceptic must admit, if in any degree a reasonable man, that there *may* exist a God, and that that God *may* have given laws to nature. No *demonstration* of the non-existence of a Great First Cause has been ever yet attempted, nor, until the knowledge of some sceptic extends over all space, ever *can* be rationally attempted. Merely to *doubt* the fact of God's existence, and to give reasons for the doubt, must till then form the highest achievements of scepticism. And the God who *may* thus exist, and who *may* have given laws to nature, *may* also have revealed himself to man, and, in order to secure man's reasonable belief in the reality of the revelation, *may* have temporarily suspended in its operation some great natural law, and have thus shown himself to be its Author and Master. Such seems to be the philosophy of miracles; which are thus evidently not only *not* impossibilities, but even not *improbabilities*. Even were we to permit the sceptic himself to fix the numbers representative of those several *mays* in the case, which I have just repeated, the chances against them, so to speak, would be less by many thousand times than the chances against the hundred dice of La Place's illustration all turning up aces. The existence of a Great First Cause is at least as probable,—the sceptic himself being judge in the matter,—as the *non*-existence of a Great First Cause; and so the probability in this first stage of the argument, instead of being, as in the case of the single die, only one to six, is as one to one. Again,—in accordance with an expectation so general among the human family as to form one of the great instincts of our nature,—an instinct to which every form of religion, true or false, bears evidence, —it is in no degree less probable that this God should have revealed himself to man, than that he should *not* have revealed himself to man ; and here the chances are again as one to one,—not, as in the second stage of the calculation on the dice, as one to thirty-six. Nor, in the third and last stage, is it less probable that God, in revealing himself to man, should have given miraculous evidence of the truth of the revelation, so that man "might believe in Him for his work's sake," than that He should *not* have done so ; and here yet again the chances are as one to one,—not as one to two hundred and sixteen. No rational sceptic could fix the chances lower ; nay, no rational sceptic, so far as the *existence* of a Great First Cause is concerned, would be inclined to fix them so low : and yet it is in order to annihilate all belief in a possibility against which the chances are so few as to be represented—scepticism itself being the actuary in the case —by three units, that Hume and La Place have framed their argument. Miracles *may* have taken place,—the probabilities against them, stated in their most extreme and exaggerated form, are by no means many or strong ; but we are nevertheless not to believe that they *did* take place, simply because miracles they were. Now, the effect of the establishment of a principle such as this would be simply, I repeat, the destruction of the ability of transmitting certain beliefs, however well founded originally, from one set or generation of men to another. These beliefs the first set or generation might, on La Place's own principles, be compelled to entertain. The evidence of the senses, however wonderful the event which they certified, is not, he himself tells us, to be resisted. But the conviction which, on one set of principles, these men were on no account to resist, the men that came immediately after them were, on quite another set of principles, on no account to entertain. And thus the anti-miracle argument, instead of leading, as all true philosophy ought, to an exact correspondence between the realities of things and the convictions received by the mind regarding them, palpably forms a bar to the reception of beliefs, adequate to the possibilities of actual occurrence or event, and so constitutes an imperfection or flaw in the mental economy, instead of working an improvement. And in accordance with this view, we find that in the economy of minds of the very highest order this imperfection or flaw has had no place. Locke studied and wrote upon the subject of miracles proper, and exhibited in his "Discourse" all the profundity of his extraordinary mind; and yet Locke was a believer. Newton studied and wrote on the subject of miracles of another kind, —those of prophecy; and he also, as shown by his "Observations on the Prophecies of Daniel and the Apocalypse," was a believer. Butler studied and wrote on the subject of miracles, chiefly in connection with "Miraculous Revelation;" and he also was a believer. Chalmers studied and wrote on the subject of miracles in his "Evidences," after Hume, La Place, and Playfair had all promulgated their peculiar views regarding it; and he also was a believer. And in none of the truly distinguished men of the present day, though all intimately acquainted with the anti-miracle argument, is this flaw or imperfection found to exist: on the contrary, they all hold, as becomes the philosophic intellect and character, that whatever is possible may occur, and that whatever occurs ought, on the proper evidence, to be believed.

But though the experience argument is of no real force, and, as shown by the beliefs of the higher order of minds, of no real effect, when brought to bear against miracles supported by the proper testimony, *it is* of great force and effect when brought to bear, not against *miracles*, but against some presumed *law*. It is experience, and experience only, that determines what is or is not law; and it is law, and law only, that constitutes the subject-matter of ordinary experience. Experience, in determining what is really miracle, does so simply through its positive knowledge of law: by knowing law, it knows also what would be a violation of it. And so miracle cannot possibly form the subject-matter of experience in the sense of Hume. For did miracle constitute the subject-matter of experience, the law of which the miracle was a violation *could not*: most emphatically, in this case, were there "no law" there could be "no transgression;" and so experience would be unable to recognise, not only the existence of the law transgressed, but also of the miracle, in its character as such, which was a transgression of the law. We determine from experience that there exists a certain fixed law, known among men as the law of gravitation; and that, in consequence of this law, if a human creature attempt standing upon the sea, he will sink into it; or if he attempt rising from the earth into the heavens, he will remain fixed to the spot on which the attempt is made. Such, in these cases, would be the direct effects of this gravitation *law;* and any presumed law antagonistic in its character could not be other than a law contrary to that invariable experience by which the existence of the real law in the case is determined. But certain it is,—for the evidence regarding the facts cannot be resisted, and by the greater minds has not been resisted,—that a man *did* once walk upon the sea without sinking into it, and *did* once ascend from the earth into the sky ; and these *miracles* ought not to be tested,—and by earnest inquirers after truth really never have been tested,—by any experience of the uniformity of the law of which they were professed transgressions, seeing it was essentially and obviously necessary that, in order to serve the great moral purpose which God intended by them, the law which they violated should have been a uniform law, and that they should have been palpable violations of it. But while the experience argument is thus of no value when directed against well-attested *miracle*, it is, as I have said, all-potent when directed against presumed *law*. Of law we know nothing, I repeat, except what experience tells us. A miracle contrary to experience in the sense of Hume is simply a miracle ; a presumed law contrary to experience is no law at all. For it is from experience, and experience only, that we know anything of natural law. The argument of Hume and La Place is perfect, as such, when directed against the development visions of the Lamarckian.

THE DEVELOPMENT HYPOTHESIS IN ITS EMBRYOTIC STATE.

OLDER THAN ITS ALLEGED FOUNDATIONS.

When Maillet first promulgated his hypothesis, many of the departments of natural history existed as mere regions of fable and romance ; and, in addressing himself to the *Muscadins* of Paris, in a popular work as wild and amusing as a fairy tale, he could safely take the liberty, and he did take it very freely, of exaggerating the marvellous, and adding fresh fictions to the untrue. And in preparing them for his theory of the metamorphoses of a marine into a terrestrial vegetation, he set himself, in accordance with his general character, to show that really the transmutation did not amount to much. "I know you have resided a long time," his Indian Philosopher is made to say, "at Marseilles. Now, you can bear me witness, that the fishermen there daily find in their nets, and among their fish, plants of a hundred kinds, with their fruits still upon them ; and though these fruits are not so large and so well nourished as those of our earth, yet the species of these plants is in no other respect dubious. They there find clusters of white and black grapes, peach-trees, pear-trees, prune-trees, apple-trees, and all sorts of flowers. When in that city, I saw, in the cabinet of a curious gentleman, a prodigious number of those sea-productions of different qualities, especially of rose-trees, which had their roses very red when they came out of the sea. I was there presented with a cluster of black sea-grapes. It was at the time of the vintage, and there were two grapes perfectly ripe."

Now, all this, and much more of the same nature, addressed to the Parisians of the reign of Louis the Fifteenth, passed, I doubt not, wonderfully well; but it will not do now, when almost every young girl, whether in town or country, is a botanist, and works on the algæ have become popular. Since Maillet wrote, Hume promulgated his argument on Miracles, and La Place his doctrine of Probabilities. There can be no doubt that these have exerted a wholesome influence on the laws of evidence; and by these laws, as restricted and amended,—laws to which, both in science and religion, we ourselves conform,—we insist on trying the Lamarckian hypothesis, and in condemning it,—should it be found to have neither standing in experience nor support from testimony,—as a mere feverish dream, incoherent in its parts and baseless in its fabric. Give, we ask, but one well-attested instance of transmutation from the algæ to even the lower forms of terrestrial vegetation common on our sea-coasts, and we will keep the question open, in expectation of more. It will not do to tell us,—as Cuvier was told, when he appealed to the fact, determined by the mummy birds and reptiles of Egypt, of the fixity of species in all, even the slightest particulars, for at least three thousand years,—that immensely extended periods of time are necessary to effect specific changes, and that human observation has not been spread over a period sufficiently ample to furnish the required data regarding them. The apology is simply a confession that, in these ages of the severe inductive philosophy, you have been dreaming your dream, cut off, as if by the state of sleep, from all the tangibilities of the real waking-day world, and that you have not a vestige of testimony with which to support your ingenious vagaries.

But on another account do we refuse to sustain the excuse. It is *not true* that human observation has not been spread over a period sufficiently extended to furnish the necessary data for testing the development hypothesis. In one special walk,—that which bears on the supposed transmutation of algæ into terrestrial plants,—human observation *has* been spread over what is strictly analogous to *millions* of years. For extent of space in this matter is exactly correspondent with duration of time. No man, in this late period of the world's history, attains to the age of five hundred years; and as some of our larger English oaks have been known to increase in bulk of trunk and extent of bough for five centuries together, no man can possibly have seen the same huge oak pass, according to Cowper, through its various stages of "treeship,"—

> "First a seedling hid in grass;
> Then twig; then sapling; and, as century rolls
> Slow after century, a giant bulk,
> Of girth enormous, with moss-cushioned root
> Upheaved above the soil, and sides embossed
> With prominent wens globose.'

But though no man lives throughout five hundred years of time, he can trace, by passing in some of the English forests through five hundred yards of space, the history of the oak in all its stages of growth, as correctly as if he *did* live throughout the five hundred years. Oaks, in the space of a few hundred yards, may be seen in every stage of growth, from the newly burst acorn, that presents to the light its two fleshy lobes, with the first tender rudiments of a leaflet between, up to the giant of the forest, in the hollow of whose trunk the red deer may shelter, and find ample room for the broad spread of his antlers. The fact of the development of the oak, from the minute two-lobed seedling of a week's growth up to the gigantic tree of five centuries, is as capable of being demonstrated by observation spread over five hundred yards of space, as by observation spread over five hundred years of time. And be it remembered, that the sea-coasts of the world are several hundred thousand miles in extent. Europe is by far the smallest of the earth's four large divisions, and it is bounded, in proportion to its size, by a greater extent of land than any of the others. And yet the sea-coasts of Europe alone, including those of its islands, exceed twenty-five thousand miles. We have results before us, in this extent of space, identical with those of many hundred thousand years of time; and if terrestrial plants were as certainly developments of the low plants of the sea as the huge oak is a development of the immature seedling just sprung from the acorn, so vast a stretch of sea-coast could not fail to present us with the intermediate vegetation in all its stages. But the sea-coasts fail to exhibit even a vestige of the intermediate vegetation. Experience spread over an extent of space analogous to millions of years of time, does not furnish, in this department, a single fact corroborative of the development theory, but, on the contrary, many hundreds of facts that bear directly against it.

The author of the "Vestiges" is evidently a practised and tasteful writer, and his work abounds in ingenious combinations of thought; but those powers of abstract reflection on whose vigorous exercise the origination of argument depends, nature seems to have denied him. There are two things in especial which his work wants,—*original observation* and *abstract thought*,—the power of *seeing* for himself and of *reasoning* for himself; and what we find instead is simply a vivid appreciation of the *images* of things, as these images exist in other minds, and a vigorous perception of the various shades of resemblance which obtain among them. There is a large amount of analogical power exhibited; but that basis of truth which correct observation can alone furnish, and that ability of nicely distinguishing differences by which the faculty of discerning similarity must be for ever regulated and governed, are wanting, in what, in a mind of fine general texture and quality, must be regarded as an extraordinary degree. And hence an ingenious but very unsolid work,—full of images transferred, not from the scientific field, but from the field of *scientific mind*, and charged with glittering but vague resemblances, stamped in the mint of fancy; which, were they to be used as mere counters in some light literary game of story-telling or character-sketching, would be in no respect out of place, but which, when passed current as the proper coin of philosophic argument, are really frauds on the popular understanding. There are, however, not a few instances in the "Vestiges" and its "Sequel," in which that defect of reflective power to which I refer rather enhances than diminishes the difficulty of reply, by presenting to the controversialist mere intangible clouds with which to grapple; that yet, through the existence of a certain superstition in the popular mind, as predisposed to accept as true whatever takes the form of science, as its predecessor the old superstition was inclined a century ago to reject science itself, are at least suited to blind and bewilder. Of this kind of difficulty, the following passage, in which the author of the work cashiers the Creator as such, and substitutes, instead, a mere animal-manufacturing piece of clock-work, which bears the name of natural law,* furnishes us with a remarkable instance.

"Admitting," he remarks, "that we see not now any such

* We are supplied with a curious example of that ever-returning cycle of speculation in which the human mind operates, by not only the introduction of the *principle* of Epicurus into the "Vestiges," but also by the unconscious employment of even his very *arguments*, slightly modified by the floating semi-scientific notions of the time. The following passages, taken, the one from the modern work, the other from Fenelon's life of the old Greek philosopher, are not unworthy of being studied, as curiously illustrative of the cycle of thought. Epicurus, I must, however, first remind the reader, in the words of his biographer, "supposed that men, and all other animals, were originally produced by the ground. According to him, the primitive earth was fat and nitrous; and the sun, gradually warming it, soon covered it with herbage and shrubs: there also began to arise on the surface of the ground a great number of small tumours like mushrooms, which having in a certain time come to maturity, the skin burst, and there came forth little animals, which, gradually retiring from the place where they were produced, began to respire." And there can be little doubt, that had the microscope been a discovery of early Greece, the passage here would have told us, not of mushroom-like tumours, but of monads. Save that the element of microscopic fact is awanting in the one and present in the other, the following are strictly parallel lines of argument:—

"To the natural objection that the earth does not now produce men, lions, and dogs, Epicurus replies that the fecundity of the earth is now exhausted. In advanced age a woman ceases to bear children; a piece of land never before cultivated produces much more during the few first years than it does afterwards; and when a forest is once cut down, the soil never produces

"In the first place, there is no reason to suppose that, though life had been imparted by natural means, after the first cooling of the surface to a suitable temperament, it would continue thereafter to be capable of being imparted in like manner. The great work of the peopling of this globe with living species is mainly a fact accomplished: the highest known species came as

fact as the production of new species, we at least know, that while such facts were occurring upon earth, there were associated phenomena in progress of a character perfectly ordinary. For example, when the earth received its first fishes, sandstone and limestone were forming in the manner exemplified a few years ago in the ingenious experiments of Sir James Hall; basaltic columns rose for the future wonder of man, according to the principle which Dr Gregory Watt showed in operation before the eyes of our fathers; and hollows in the igneous rocks were filled with crystals, precisely as they could now

---

trees equal to those which have been rooted up. Those which are afterwards planted become dwarfish, and are perpetually degenerating. We are, however, he argues, by no means certain but there may be at present rabbits, hares, foxes, bears, and other animals, produced by the earth in their perfect state. The reason why we are backward in admitting it is, that it happens in retired places, and never falls under our view; and, never seeing rats but such as have been produced by other rats, we adopt the opinion that the earth never produced any." (*Fenelon's Lives of the Ancient Philosophers.*)

a crowning effort thousands of years ago. The work being thus to all appearance finished, we are not necessarily to expect that the origination of life and of species should be conspicuously exemplified in the present day. We are rather to expect that the vital phenomena presented to our eyes should mainly, if not entirely, be limited to a regular and unvarying succession of races by the ordinary means of generation. This, however, is no more an argument against a time when phenomena of the first kind prevailed, than it would be a proof against the fact of a mature man having once been a growing youth, that he is now seen growing no longer. * * * Secondly, it is far from being certain that the primitive imparting of life and form to inorganic elements is not a fact of our times. (*Vestiges of Creation.*)

be by virtue of electric action, as shown within the last few years by Crosse and Becquerel. The seas obeyed the impulse of gentle breezes, and rippled their sandy bottoms, as seas of the present day are doing; the trees grew as now, by favour of sun and wind, thriving in good seasons and pining in bad: this while the animals above fishes were yet to be created. The movements of the sea, the meteorological agencies, the disposition which we see in the generality of plants to thrive when heat and moisture were most abundant, were kept up in silent serenity, as matters of simply natural order, throughout the whole of the ages which saw reptiles enter in their various forms upon the sea and land. It was about the time of the first mammals that the forest of the Dirt-Bed was sinking in natural ruin amidst the sea sludge, as forests of the Plantagenets have been doing for several centuries upon the coast of England. In short, *all the common operations of the physical world were going on in their usual simplicity, obeying that order which we still see governing them;* while the supposed extraordinary causes were in requisition for the development of the animal and vegetable kingdoms. There surely hence arises a strong presumption against any such causes. It becomes much more likely that the latter phenomena were evolved in the manner of law also, and that we only dream of extraordinary causes here, as men once dreamt of a special action of Deity in every change of wind and the results of each season, merely because they did not know the laws by which the events in question were evolved."

How, let us suppose, would David Hume,—the greatest thinker of which infidelity can boast,—have greeted the auxiliary who could have brought him such an *argument* as a contribution to the cause? "Your objection, so far as you have stated it," the philosopher might have said, "amounts simply to this:—Creation by direct act is a miracle; whereas all that exists is *propagated* and *maintained* by natural law. Natural laws,—to vary the illustration,—were in full operation at the period when the Author of the Christian religion was, it is said, engaged in working his miracles. When, according to our opponents, he walked upon the surface of the sea, Peter, through the operation of the natural law of gravitation, was sinking into it; when he withered, by a word, the barren fig-tree, there were other trees on the Mount thriving in conformity with the vegetative laws, under the influence of sun and shower; when he raised the dead Lazarus, there were corpses in the neighbouring tombs passing, through the natural putrefactive fermentation, into a state of utter decomposition. In fine, at the time when he was engaged, as Reid and Campbell believe, in working miracles in violation of law, the laws of which these were a violation actually existed, and were everywhere actively operative; or, to employ your own words, when the New Testament miracles were, it is alleged, in the act of being wrought, 'all the common operations of the physical world were going on in their usual simplicity, obeying that order which we still see governing them.' Such is the portion of your statement already made; what next?" "It is surely very unlikely," replies the auxiliary, "that in such a complex mass of phenomena there should have been two totally distinct modes of the exercise of the Divine power,—the mode by miracle and the mode by law." "Unlikely!" rejoins the philosopher; "on what grounds?" "O, just *unlikely*," says the auxiliary;—"unlikely that God should be at once operating on matter through the agency of natural laws, of which *man knows much*, and through the agency of miraculous acts, of the nature of which *man knows nothing*. But I have not thought out the subject any farther : you have, in the statement already made, my entire *argument*." "Ay, I see," the author of the "Essay on Miracles" would probably have remarked; "you deem it unlikely that Deity should not only work in part, as he has always done, by means of which *men*, —clever fellows like you and me,—think they know a great deal, but that he should also work in part, *as he has always done*, by means of which they know nothing at all. Admirably reasoned out! You are, I make no doubt, a sound, zealous unbeliever in your private capacity, and your argument may have great weight with your own mind, and be, in consequence, worthy of encouragement in a small way ; but allow me to suggest that, for the sake of the general cause, it should be kept out of reach of the enemy. There are in the Churches Militant on both sides of the Tweed shrewd combatants, who have nearly as much wit as ourselves." I think I understand the reference of the author of the "Vestiges" to the *dream* "of a special action of Deity in every change of wind and the results of each season." Taken with what immediately goes before, it means something considerably different from those fancies of the "untutored Indian," who, according to the poet,

> "Sees God in clouds, or hears him in the wind."

There is a school of infidelity, tolerably well known in the capital of Scotland as by far the most superficial which our country has yet seen, that measures mind with a tape-line and the callipers, and, albeit not Christian, laudably exemplifies, in a loudly expressed regard for science, the Christian grace of loving its enemy. And the belief in a special Providence, who watches over and orders all things, and without whose permission there falleth not even a "sparrow to the ground," the apostles of this school set wholly aside, substituting, instead, a belief in the indiscriminating operation of natural laws ; as if, with the broad fact before them that even man can work out his will merely by knowing and directing these laws, the God by whom they were instituted should lack either the power or the wisdom to make them the pliant ministers of *his*. It is, I fear, to the distinctive tenet in the creed of this hapless school that the author of the "Vestiges" refers. Nor is it in the least surprising, that a writer who labours through two carefully written volumes,* to destroy the existing belief in "God's works of Creation," should affect to hold that the belief in his "works of Providence" had been destroyed already. But faith in a special superintendence of Deity is not yet dead : nay, more, He who created the human mind took especial care, in its construction, that, save in a few defective specimens of the race, the belief should never die.

The author of the "Vestiges" complains of the illiberality with which he has been treated. "It has appeared to various critics," we find him saying, "that very sacred principles are threatened by a doctrine of universal law. A natural origin of life, and a natural basis in organization for the opèrations of the human mind, speak to them of fatalism and materialism. And, strange to say, those who every day give views of *physical cosmogony* altogether discrepant in appearance with that of Moses, apply hard names to my book for suggesting an *organic cosmogony* in the same way, liable to inconsiderate odium. I must firmly protest against this mode of meeting speculations regarding nature. The object of my

* "*Vestiges of the Natural History of Creation*," and "*Explanations, being a Sequel to the Vestiges.*"

book, whatever may be said of the manner in which it is treated, is purely scientific. The views which I give of the history of organization stand exactly on the same ground upon which the geological doctrines stood fifty years ago. I am merely endeavouring to read aright another chapter of the mystic book which God has placed under the attention of his creatures. . . The absence of all liberality in my reviewers is striking, and especially so in those whose geological doctrines have exposed them to similar misconstruction. If the men newly emerged from the odium which was thrown upon Newton's theory of the planetary motions, had rushed forward to turn that odium upon the patrons of the dawning science of Geology, they would have been prefiguring the conduct of several of my critics, themselves hardly escaped from the rude hands of the narrow-minded, yet eager to join that rabble against a new and equally unfriended stranger, as if such were the best means of purchasing impunity for themselves. *I trust that a little time will enable the public to penetrate this policy.*"

Now, there is one very important point to which the author of this complaint does not seem to have adverted. The astronomer founded his belief in the mobility of the earth and the immobility of the sun, not on a mere dream-like hypothesis, founded on nothing, but on a wide and solid base of pure induction. Galileo was no mere dreamer ;—he was a discoverer of great truths, and a profound reasoner regarding them : and on his discoveries and his reasonings, compelled by the inexorable laws of his mental constitution, did he build up certain deductive beliefs, which had no previous existence in his mind. His convictions were consequents, not antecedents. Such, also, is the character of geological discovery and inference, and of the existing belief,—their joint production,—regarding the great antiquity of the globe. No geologist worthy of the name *began* with the belief, and then set himself to square geological phenomena with its requirements. It is a deduction,—a result ;—not the starting assumption, or given sum, in a process of calculation, but its ultimate finding or answer. Clergymen of the orthodox Churches, such as the Sumners, Sedgwicks, Bucklands, Conybeares, and Pye Smiths of England, or the Chalmerses, Duncans, and Flemings of our own country, must have come to the study of this question of the world's age with at least no bias in favour of the geological estimate. The old, and, as it has proven, erroneous reading of the Mosaic account, was by much too general a one early in the present century, not to have exerted upon them, in their character as ministers of religion, a sensible influence of a directly opposite nature. And the fact of the complete reversal of their original bias, and of the broad unhesitating finding on the subject which they ultimately substituted instead, serves to intimate to the uninitiated the strength of the evidence to which they submitted. There can be nothing more certain than that it is minds of the same calibre and class, engaged in the same inductive track, that yielded in the first instance to the astronomical evidence regarding the earth's motion, and, in the second, to the geological evidence regarding the earth's age.*

---

* The chapter in which this passage occurs originally appeared, with several of the others, in the *Witness* newspaper, in a series of articles, entitled "Rambles of a Geologist," and drew forth the following letter from a correspondent of the *Scottish Press,* the organ of a powerful and thoroughly respectable section of the old Dissenters of Scotland. I present it to the reader merely to show, that if, according to the author of the "Vestiges," geologists assailed the development hypothesis in the fond hope of "purchas-

But how very different the nature and history of the development hypothesis, and the character of the intellects with whom it originated, or by whom it has been since

---

ing impunity for themselves," they would succeed in securing only disappointment for their pains :—

"THE PRE-ADAMITE EARTH.

"*To the Editor of the Scottish Press.*

"SIR,—I occasionally observe articles in your neighbour and contemporary the *Witness*, characteristically headed 'Rambles of a Geologist,' wherein the writer with great zeal once more 'slays the slain' heresies of the 'Vestiges of Creation.' This writer (of the 'Rambles,' I mean), nevertheless, and at the same time, announces his own tenets to be much of the same sort as applied to mere dead matter, that those of the 'Vestiges' are with regard to living organisms. He maintains that the world, during the last million of years, has been of itself rising or developing, without the interposition of a miracle, from chaos into its present state; and, of course, as it is still, as a world, confessedly far below the acme of physical perfection, that it must be just now on its passage, self-progressing, towards that point, which terminus it may reach in another million of years hence [!!!] The author of the 'Vestiges,' as quoted by the author of the 'Rambles,' in the last number of the *Witness*, complains that the latter and his allies are not at all so liberal to him as, from their present circumstances and position, he had a right to expect. He (the author of the 'Vestiges'), reminds his opponents that they have themselves only lately emerged from the antiquated scriptural notions that our world was the direct and almost immediate construction of its Creator,—as much so, in fact, as any of its organized tenants,—and that it was then created in a state of physical excellence, the highest possible, to render it a suitable habitation for these tenants, and all this only about six or seven thousand years ago, —to the new light of their present *physico-Lamarckian* views; and he asks, and certainly not without reason, why should *these men*, so circumstanced, be so anxious to stop him in his attempt to move one step farther forward in the very direction they themselves have made the last move?—that is, in his endeavour to ex-

adopted! In the first place, it existed as a wild dream ere Geology had any being as a science. It was an antecedent, not a consequent,—a starting assumption, not a result. No

---

tend their own principles of self-development from mere matter to living creatures. Now, Sir, I confess myself to be one of those (and possibly you may have more readers similarly constituted) who not only cannot see any great difference between merely *physical* and *organic* development [!!] but who would be inclined to allow the latter, absurd as it is, the advantage in point of likelihood [!!!] The author of the 'Rambles,' however, in the face of this, assures us that *his* views of physical self-development and long chronology belong to the inductive sciences. Now, I could at this stage of his rambles have wished very much that, instead of merely *saying* so, he had given his *demonstration.* He refers, indeed, to several great men, who, he says, are of his opinions. Most that these men have written on the question at issue I have seen, but it appeared far from demonstrative, and some of them, I know, had not fully made up their mind on the point [!!!] Perhaps the author of the 'Rambles' could favour us with the inductive process that converted himself; and, as the attainment of truth, and not victory, is my object, I promise either to acquiesce in or rationally refute it [?] Till then I hold by my antiquated tenets, that our world, nay, the whole material universe, was created about six or seven thousand years ago, and that in a state of physical excellence of which we have in our present fallen world only the 'vestiges of creation.' I conclude by mentioning that this view I have held now for nearly thirty years, and, amidst all the vicissitudes of the philosophical world during that period, I have never seen cause to change it. Of course, with this view I was, during the interval referred to, a constant opponent of the once famous, though now exploded, nebular hypothesis of La Place; and I yet expect to see *physical development* and *long chronology* wither also on this earth, now that THEIR ROOT (the said hypothesis) has been eradicated from the SKY[!!!]—I am, Sir, your most obedient servant,

"PHILALETHES."

I am afraid there is little hope of converting a man who has held so stoutly by his notions "for nearly thirty years;" especially as, during that period, he has been acquainting himself with

one will contend that Maillet was a geologist. Geology had no place among the sciences in the age in which he lived, and even no name. And yet there is a translation of his

---

what writers such as Drs Chalmers, Buckland, and Pye Smith have written on the other side. But for the *demonstration* which he asks, as *I* have conducted it, I beg leave to refer him to the seventeenth chapter of my little work, "First Impressions of England and its People." I am, however, inclined to suspect that he is one of a class whose objections are destined to be removed rather by the operation of the laws of matter than of those of mind. For it is a comfortable consideration, that in this controversy the geologists *have* the laws of matter on their side;—"the stars in their courses fight against Sisera." Their opponents now, like the opponents of the astronomer in the ages gone by, are, in most instances, men who have been studying the matter "for nearly thirty years." When they study it for a few years longer they disappear; and the men of the same cast and calibre who succeed them are exactly the men who throw themselves most confidently into the arms of the enemy, and look down upon their poor silent predecessors with the loftiest commiseration. It is, however, not uninstructive to remark how thoroughly, in some instances, the weaker friends and the wilier enemies of Revelation are at one in their conclusions respecting natural phenomena. The correspondent of the *Scottish Press* merely regards the views of the author of the "Vestiges" as possessing "the advantage, in point of likelihood," over those of the geologists his antagonists: his ally the Dean of York goes greatly farther, and stands up as stoutly for the transmutation of species as Lamarck himself. Descanting, in his *New System of Geology*, on the various forms of trilobites, ammonites, belemnites, &c. Dean Cockburn says,—

"These creatures appear to have possessed the power of secreting from the stone beneath them a limy covering for their backs, and, perhaps, fed partly on the same solid material. Supposing, now, that the first trilobites were destroyed by the Llandeilo Slates, some spawn of these creatures would arise above these flags, and, after a time, would be warmed into existence. These *molluscs* [!!] then, having a better material from which to extract their food and covering, would probably expand in a slightly different form, and with a more extensive mantle than what belonged to the

Telliamed now lying before me, bearing date 1750, in which I find very nearly the same account given of the origin of

---

parent species. The same would be still more the case with a new generation, fed upon a new deposit from some deeper volcano, such as the Caradoc or Wenlock Limestone, in which lime more and more predominates. Now, if any one will examine the various prints of trilobites in Sir R. Murchison's valuable work, he will find but very trifling differences in any of them [!!] and those differences only in the stony covering of their backs. I knew two brothers once much alike: the one became a curate with a large family; the other a London alderman. If the skins of these two pachydermata had been preserved in a fossil state, there would have been less resemblance between them than between an *asaphus tyrannus* and an *asaphus caudatus*. * * * A careful and laborious investigation has discovered, as in the trilobites, a difference in the ammonites of different strata; but such differences, as in the former case, exist only in the form of the external shell, and may be explained in the same manner [!!] * * * As to the scaphites, baculites, belemnites, and all the other *ites* which learned ingenuity has so named, you find them in various strata the same in all important particulars, but also differing slightly in their outward coverings, as might be expected from the different circumstances in which each variety was placed [!!] The sheep in the warm valleys of Andalusia have a fine covering like to hair; but remove them to a northern climate, and in a few generations the back is covered with shaggy wool. The animal is the same,—the covering only is changed. * * * The learned have classed those shells under the names of terebratula, orthis, atrypha, pecten, &c. They are all much alike [!!!] It requires an experienced eye to distinguish them one from another: what little differences have been pointed out may readily be ascribed, as before, to difference of situation" [!!!]

The author of the "Vestiges," with this, the fundamental portion of his case, granted to him by the Dean, will have exceedingly little difficulty in making out the rest for himself. The passage is, however, not without its value, as illustrative of the darkness, in matters of physical science, "even darkness which may be felt," that is suffered to linger, in this the most scientific of ages, in the Church of Buckland, Sedgwick, and Conybeare.

animals and plants as that in the "Vestiges," and in which the sea is described as that great and fruitful womb of nature in which organization and life first began. Lamarck, at the time when Maillet wrote, was a boy in his sixth year. He became, comparatively early in life, a skilful botanist and conchologist; but not until turned of fifty did he set himself to study general zoology; and his greater work on the invertebrate animals, on which his fame as a naturalist chiefly rests, did not *begin* to appear,—for it was published serially,—until the year 1815. But his development hypothesis, identical with that of the "Vestiges," was given to the world long before,—in 1802; at a time when it had not been ascertained that there existed placoids during the Silurian period, or ganoids during the Old Red Sandstone period, or enaliosaurs during the Oolitic period; and when, though Smith had constructed his "Tabular View of the British Strata," his map had not yet appeared, and there was little more known regarding the laws of superposition among the stratified rocks than was to be found in the writings of Werner. And if the presumption be strong, in the circumstances, that Lamarck originated his development hypothesis ere he became in any very great degree skilful as a zoologist, it is no mere presumption, but a demonstrable truth, that he originated it ere he became a geologist; for a geologist he never became. In common with Maillet and Buffon, he held by Leibnitz's theory of a universal ocean; and such, as we have already seen, was his ignorance of fossils, that he erected dermal fragments of the Russian *Asterolepis* into a new genus of Polyparia,—an error into which the merest tyro in palæontology could not now fall. Such, in relation to these sciences, was the man who perfected the dream of development. Nor has the most distinguished of its continental assertors now living,—Professor Oken,—any higher claim to be regarded as a disciple of the inductive school of Geology than Lamarck. In the preface to the recently-published translation of his "Physio-Philosophy," we find the following curious confession :—"I wrote the first edition of 1810 *in a kind of inspiration,* and on that account it was not so well arranged as a systematic work ought to be. Now, though this may appear to have been amended in the second and third edition, yet still it was not possible for me to completely attain the object held in view. The book has therefore remained essentially the same as regards its fundamental principles. It is only the empirical arrangement into series of plants and animals that has been modified from time to time, *in accordance with the scientific elevation of their several departments,* or *just as discoveries and anatomical investigations have increased, and rendered some other position of the objects a matter of necessity.*" An interesting piece of evidence this ; but certainly rather simple as a confession. It will be found that while whatever gives value to the "Physio-Philosophy" of the German Professor (a work which, if divested of all the inspired bits, would be really a good one), was acquired either before or since its first appearance in the ordinary way, its development hypothesis came direct from the god. Further, as I have already had occasion to state, Oken holds, like Lamarck and Maillet, by the universal ocean of Leibnitz : he holds also, that the globe is a vast crystal, just a little flawed in the facets; and that the three granitic components,—quartz, feldspar, and mica,—are simply the hail-drops of heavy stone showers that shot athwart the original ocean, and accumulated into rock at the bottom, as snow or hail shoots athwart the upper atmosphere, and accumulates, in the form of ice, on the summits of high hills, or in the arctic or antarctic regions.

Such, in the present day, are the geological notions of Oken! They were doubtless all promulgated in what is modestly enough termed "a *kind* of inspiration;" and there are few now so ignorant of Geology as not to know that the *possessing* agent in the case,—for *inspiration* is not quite the proper word,—must have been at least of kin to that ingenious personage who volunteered of old to be a lying spirit in the mouths of the four hundred prophets. And the well-known fact, that the most popular contemporary expounder of Oken's hypothesis,—the author of the "Vestiges,"—has in every edition of his work been correcting, modifying, or altogether withdrawing his statements regarding both geological and zoological phenomena, and that his gradual development as a geologist and zoologist, from the sufficiently low type of acquirement to which his first edition bore witness, may be traced, in consequence, with a distinctness and certainty which we in vain seek in the cases of presumed development which he would so fain establish,—has in its bearing exactly the same effect. His development hypothesis was complete at a time when his geology and zoology were rudimental and imperfect. Give me your facts, said the Frenchman, that I may accommodate them to my theory. And no one can look at the progress of the Lamarckian hypothesis, with reference to the dates when, and the men by whom, it was promulgated, without recognising in it one of perhaps the most striking embodiments of the Frenchman's principle which the world ever saw. It is not the illiberal religionist that rejects and casts it off,—it is the inductive philosopher. Science addresses its assertors in the language of the possessed to the sons of Sceva the Jew:—"The astronomer I know, and the geologist I know; but who are ye?"

One of the strangest passages in the "Sequel to the Vestiges," is that in which its author carries his appeal from the tribunal of science to "another tribunal," indicated but not named, before which "this new philosophy" [remarkable chiefly for being neither philosophy nor new] "is to be truly and righteously judged." The principle is obvious, on which, were his opponents mere theologians, wholly unable, though they saw the mischievous character and tendency of his conclusions, to disprove them scientifically, he might appeal from theology to science: "it is with scientific truth," he might urge, "not with moral consequences, that I have aught to do." But on what allowable principle, professing, as he does, to found his theory on scientific fact, can he appeal from science to the want of it? "After discussing," he says, "the whole arguments on both sides in so ample a manner, it may be hardly necessary to advert to the objection arising from the mere fact, that nearly all the scientific men are opposed to the theory of the 'Vestiges.' As this objection, however, is likely to be of some avail with many minds, it ought not to be entirely passed over. If I did not think there were reasons, independent of judgment, for the scientific class coming so generally to this conclusion, I might feel the more embarrassed in presenting myself in direct opposition to so many men possessing talents and information. As the case really stands, the ability of this class to give at the present a true response upon such a subject appears extremely challengeable. It is no discredit to them that they are, almost without exception, engaged each in his own little department of science, and able to give little or no attention to other parts of that vast field. From year to year, and from age to age, we see them at work, adding, no doubt, much to the known, and advancing many important interests, but at the same time doing little for the establishment of comprehensive views of nature. Experiments in however narrow a walk, facts of whatever minuteness, make reputations in scientific societies: all beyond is regarded with suspicion and distrust. The consequence is, that philosophy, as it exists amongst us, does nothing to raise its votaries above the common ideas of their time. There can therefore be nothing more conclusive against our hypothesis in the disfavour of the scientific class, than in that of any other section of educated men."

This is surely a very strange statement. Waiving altogether the *general* fact, that great original discoverers in any department of knowledge are never men of one science or one faculty, but possess, on the contrary, breadth of mind and multiplicity of acquirement;—waiving, too, the *particular* fact, that the more distinguished original discoverers of the present day rank among at once its most philosophic, most elegant, and most extensively informed writers;—granting, for the argument's sake, that our scientific men *are* men of narrow acquirement, and "exclusively engaged, each in his own little department of science;"—it is surely rational to hold, notwithstanding, that in at least these little departments they have a better right to be heard than any other class of persons whatever. We must surely not refuse to the man of science what we at once grant to the common mechanic. A cotton-weaver or calico-printer may be a very narrow man, "exclusively engaged in his own little department;" and yet certain it is that, in a question of cotton-weaving or of calico-printing, his evidence is justly deemed more conclusive in courts of law than that of any other man, however much his superior in general breadth and intelligence. And had the author of the "Vestiges" founded his hypothesis on certain facts pertaining to the arts of cotton-weaving and calico-printing, the cotton-weaver and calico-printer would have an indisputable right to be heard on the question of their general correctness. Are we to regard the case as different because it is on facts pertaining to science, not to cotton-weaving or calico-printing, that he professes to found? His hypothesis, unless supported by scientific evidence, is a mere dream,—a fiction as baseless and wild as any in the "Fairy Tales" or the "Arabian Nights." And, fully sensible of the fact, he calls in as witnesses the physical sciences, and professes to take down their evidence. He calls into court Astronomy, Geology, Phytology, and Zoology. "Hold!" exclaims the astronomer, as the examination goes on; "you are taking the evidence of my special science most unfairly: I challenge a right of cross-examining the witness." "Hold!" cries the geologist; you are putting my science to the question, and extorting from it, in its agony, a whole series of fictions: I claim the right of examining it fairly and softly, and getting from it just the sober truth, and nothing more." And the phytologist and zoologist urge exactly similar claims. "No, gentlemen," replies the author of the "Vestiges," "you are narrow men, confined each of you to his own little department, and so I will not permit you to cross-examine the witnesses." "What!" rejoin the men of science, "not permit us to examine our own witnesses!—refuse to us what you would at once concede to the cotton-weaver or the calico-printer, were the question one of cotton-weaving or of calico-printing! We are surely not much narrower men than the man of cotton or the man of calico. It is but in our own little departments that we ask to be heard." "But you shall not be heard, gentlemen," says the author of the "Vestiges;" "at all events, I shall not care one farthing for anything you say. For observe, gentlemen, my hypothesis is nothing without the evidence of your sciences; and you all unite, I see, in taking that evidence from me; and so I confidently raise my appeal in this matter to people who know nothing about either you or your sciences. It must be before another tribunal that the new philosophy is to be truly and righteously judged." Alas! what can this mean? or where are we to seek for that tribunal of last resort to which this ingenious man refers with such confidence the consideration of his case? Can it mean, that he appeals from the only class of persons qualified to judge of his facts, to a class ignorant of these, but disposed by habits of previous scepticism to acquiesce in his conclusions, and take his premises for granted;—that he appeals from astronomers and geologists to low-minded materialists and shallow phrenologers,—from phytologists and zoologists to mesmerists and phreno-mesmerists?

I remember being much struck, several years ago, by a remark dropped in conversation by the late Rev. Mr Stewart of Cromarty, one of the most original-minded men I ever knew. "In reading in my Greek New Testament this morning," he said, "I was curiously impressed by a thought which, simple as it may seem, never occurred to me before. The portion which I perused was in the First Epistle of Peter; and as I passed from the thinking of the passage to the language in which it is expressed,—'This Greek of the untaught Galilean fisherman,'—I said, 'so admired by scholars and critics for its unaffected dignity and force, was not acquired, as that of Paul may have been, in the ordinary way, but formed a portion of the Pentecostal gift! Here, then, immediately under my eye, on these pages, are there embodied, not, as in many other parts of the Scriptures, the mere *details* of a miracle, but the direct *results* of a miracle. How strange! Had the old tables of stone been placed before me, with what an awestruck feeling would I have looked on the characters traced upon them by God's own finger! How is it that I have failed to remember that, in the language of these Epistles, miraculously impressed by the Divine power upon the mind, I possessed as significant and suggestive a relic as that which the inscription miraculously impressed by the Divine power upon the stone could possibly have furnished?'" It was a striking thought; and in the course of our walk, which led us over richly fossiliferous beds of the Old Red Sandstone, to a deposit of the Eathie Lias, largely charged with the characteristic remains of that formation, I ventured to connect it with another. "In either case," I remarked, as we seated ourselves beside a sea-cliff, sculptured over with the impressions of extinct plants and shells, "your relics, whether of the Pentecostal Greek or of the characters inscribed on the old tables of stone, could address themselves to but previously existing belief. The sceptic would see in the Sinaitic characters, were they placed before him, merely the work of an ordinary tool; and in the Greek of Peter and John, a well-known language, acquired, he would hold, in the common way. But what say you to the relics that stand out in such bold relief from the rocks beside us, in *their* character as the results of miracle? The perished tribes and races which they represent all *began* to exist. There is no truth which science can more conclusively demonstrate than that they had all a beginning. The infidel who, in this late age of the world, would attempt falling back on the fiction of an 'infinite series,' would be laughed to scorn. They all began to be. But how? No true geologist holds by the development hypothesis;—it has been resigned to sciolists and smatterers;—and there is but one other alternative. They began to be, *through the miracle of creation.* From the evidence furnished by these rocks we are shut down either to the belief in *miracle*, or to the belief in something else infinitely harder of reception, and as thoroughly unsupported by testimony as it is contrary to experience. Hume is at length answered by the severe truths of the stony science. He was not, according to Job, 'in league with the stones of the field,' and they have risen in irresistible warfare against him in the Creator's behalf."

FINAL CAUSES.—THEIR BEARING ON GEOLOGIC HISTORY.

CONCLUSION.

"NATURAL History has a principle on which to reason," says Cuvier, "which is peculiar to it, and which it employs advantageously on many occasions: it is that of the *conditions of existence*, commonly termed *final causes.*"

In Geology, which is Natural History extended over all ages, this principle has a still wider scope,—embracing not merely the characteristics and conditions of the beings which now exist, but of all, so far as we can learn regarding them, which have ever existed,—and involving the consideration of not merely their peculiarities as races placed before us without relation to time, but also of the history of their rise, increase, decline, and extinction. In studying the *biography*, if I may so express myself, of an individual animal, we have to acquaint ourselves with the circumstances in which nature has placed it,—its adaptation to these, both in structure and instinct,—the points of resemblance which it presents to the individuals of other races and families,—and the laws which determine its terms of development, vigorous existence, and decay. And all Natural History, when restricted to the passing *now* of the world's annals, is simply a congeries of biographies. It is when we extend our view into the geological field that it passes from *biography* into *history proper*, and that we have to rise from the consideration of the birth and death of individuals, which, in all mere biographies, form the great terminal events that constitute beginning and end, to a survey of the birth and death of races, and the elevation or degradation of dynasties and sub-kingdoms.

We learn from human history that nations are as certainly mortal as men. They enjoy a greatly longer term of existence, but they die at last: Rollin's History of Ancient Nations is a history of the dead. And we are taught by geological history, in like manner, that *species* are as mortal as individuals and nations, and that even genera and families become extinct. There is no *man* upon earth at the present moment whose age greatly exceeds an hundred years;—there is no *nation* now upon earth (if we perhaps except the long-lived Chinese) that also flourished three thousand years ago;—there is no *species* now living upon earth that dates beyond the times of the Tertiary deposits. All bear the stamp of death,—individuals,—nations,—species; and we may scarce less safely predicate, looking upon the past, that it is appointed for nations and species to die, than that it "is appointed for *man* once to die." Even our own species, *as now constituted*,—with instincts that conform to the original injunction, "increase and multiply," and that, in consequence, "marry and are given in marriage,"—shall one day cease to exist: a fact not less in accordance with beliefs inseparable from the faith of the Christian, than with the widely founded experience of the geologist. Now, it is scarce possible for the human mind to become acquainted with the fact, that at certain periods species began to exist, and then, after the lapse of untold ages, ceased to be, without enquiring whether, from the "conditions of existence, commonly termed final causes," we cannot deduce a reason for their rise or decline, or why their term of being should have been included rather in one certain period of time than another. The same faculty which finds employment in tracing to their causes the rise and fall of nations, and which it is the merit of the philosophic historian judiciously to exercise, will to a certainty seek employment in this department of history also; and that there will be an appetency for such speculations in the public mind, we may infer from the success, as a literary undertaking, of the "Vestiges of Creation,"—a work that bears the same sort of relation in this special field to sober enquiry, founded on the true conditions of things, that the legends of the old chroniclers bore to authentic history. The progressive state of geologic science has hitherto militated against the formation of theory of the soberer character. Its facts,—still merely in the forming,—are necessarily imperfect in their classification, and limited in their amount; and thus the essential data continues incomplete. Besides, the men best acquainted with the basis of fact which already exists have quite enough to engage them in adding to it. But there are limits to the field of palæontological discovery, in its relation to what may be termed the chronology of organized existence, which, judging from the progress of the science in the past, may be well nigh reached in favoured localities, such as the British islands, in about a quarter of a century from the present time; and then, I doubt not, geological history, in legitimate conformity with the laws of mind, and from the existence of the pregnant principle peculiar, according to Cuvier, to that science of which Geology is simply an extension, will assume a very extraordinary form. We cannot yet aspire "to the height of this great argument:"

our foundations are in parts still unconsolidated and incomplete, and unfitted to sustain the perfect superstructure which shall one day assuredly rise upon them; but from the little which we can now see, "as if in a glass darkly," enough appears from which to

> "Assert eternal Providence,
> And justify the ways of God to men."

The history of the four great monarchies of the world was typified, in the prophetic dream of the ancient Babylonish king, by a colossal image, "terrible in its form and brightness," of which the "head was pure gold," the "breast and arms of silver," the "belly and thighs of brass," and the legs and feet "of iron, and of iron mingled with clay." The vision in which it formed the central object was appropriately that of a puissant monarch; and the image itself typified the merely human monarchies of the earth. It would require a widely different figure to symbolize the great monarchies of creation. And yet Revelation does furnish such a figure. It is that which was witnessed by the captive prophet beside "the river Chebar," when "the heavens were opened, and he saw visions of God." In that chariot of Deity, glowing in fire and amber, with its complex wheels "so high that they were dreadful," set round about with eyes, there were living creatures, of whose four faces three were brute and one human; and high over all sat the Son of Man. It would almost seem as if, in this sublime vision,—in which, with features distinct enough to impress the imagination, there mingle the elements of an awful incomprehensibility, and which even the genius of Raffaelle has failed adequately to portray,—the history of all the past and of all the future had been symbolized. In the order of Providence intimated in the geologic record, the brute faces, as in the vision, outnumber the human ;—the human dynasty is one, and the dynasties of the inferior animals are three ; and yet who can doubt that they all equally compose parts of a well-ordered and perfect whole, as the four faces formed but one cherubim ; that they have been moving onward to a definite goal, in the unity of one grand harmonious design,—now " lifted up high " over the comprehension of earth,—now let down to its humble level ; and that the Creator of all has been ever seated over them on the throne of his providence,—a " likeness in the appearance of a man,"—embodying the perfection of his nature in his workings, and determining the end from the beginning ?

There is geologic evidence, as has been shown, that in the course of creation the higher orders succeeded the lower. We have no good reason to believe that the mollusc and crustacean preceded the fish, seeing that discovery, in its slow course, has already traced the vertebrata in the ichthyic form, down to deposits which only a few years ago were regarded as representative of the first beginnings of organized existence on our planet, and that it has at the same time failed to add a lower system to that in which their remains occur. But the fish seems most certainly to have preceded the reptile and the bird ; the reptile and the bird to have preceded the mammiferous quadruped ; and the mammiferous quadruped to have preceded man,—rational, accountable man, whom God created in his own image,—the much-loved Benjamin of the family,—last-born of all creatures. It is of itself an extraordinary fact, without reference to other considerations, that the order adopted by Cuvier, in his animal kingdom, as that in which the four great classes of vertebrate animals, when marshalled according to their rank and standing, naturally range, should be also that in which they occur in order of time. The brain which bears an average proportion to the spinal cord of not more than two to one, came first,—it is the brain of the fish; that which bears to the spinal cord an average proportion of two and a half to one succeeded it,—it is the brain of the reptile; then came the brain averaging as three to one,—it is that of the bird; next in succession came the brain that averages as four to one,—it is that of the mammal; and last of all there appeared a brain that averages as *twenty-three* to one,—reasoning, calculating man had come upon the scene. All the facts of geological science are hostile to the Lamarckian conclusion, that the lower brains were developed into the higher. As if with the express intention of preventing so gross a mis-reading of the record, we find, in at least two classes of animals,—fishes and reptiles,—the higher races placed at the beginning: the slope of the inclined plane is laid, if one may so speak, in the reverse way, and, instead of rising towards the level of the succeeding class, inclines downwards, with at least the effect, if not the design, of making the break where they meet exceedingly well marked and conspicuous. And yet the record does seem to speak of *development and progression*;—not, however, in the province of organized existence, but in that of insensate matter, subject to the purely chemical laws. It is in the style and character of *the dwelling-place* that gradual improvement seems to have taken place,—not in the functions or the rank of any class of its inhabitants; and it is with special reference to this gradual improvement in our common mansion-house the earth, in its bearing on the "conditions of existence," that not a few of our reasonings regarding the introduction and extinction of species and genera must proceed.

That definite period at which man was introduced upon the scene seems to have been specially determined by the conditions of correspondence which the phenomena of his habitation had at length come to assume with the predestined constitution of his mind. The large reasoning brain would have been wholly out of place in the earlier ages. It is indubitably the nature of man to base the conclusions which regulate all his actions on fixed phenomena ;—he reasons from cause to effect, or from effect to cause ; and when placed in circumstances in which, from some lack of the necessary basis, he cannot so reason, he becomes a wretched, timid, superstitious creature, greatly more helpless and abject than even the inferior animals. This unhappy state is strikingly exemplified by that deep and peculiar impression made on the mind by a severe earthquake, which Humboldt, from his own experience, so powerfully describes. "This impression," he says, "is not, in my opinion, the result of a recollection of those fearful pictures of devastation presented to our imagination by the historical narratives of the past, but is rather due to the sudden revelation of the delusive nature of the inherent faith by which we had clung to a belief in the immobility of the solid parts of the earth. We are accustomed from early childhood to draw a contrast between the mobility of water and the immobility of the soil on which we tread ; and this feeling is confirmed by the evidence of our senses. When, therefore, we suddenly feel the ground move beneath us, a mysterious force, with which we were previously unacquainted, is revealed to us as an active disturber of stability. A moment destroys the illusion of a whole life ; our deceptive faith in the repose of nature vanishes ; and we feel transported into a realm of unknown destructive forces. Every sound,—the faintest motion of the air,—arrests our attention, and we no longer trust the ground on which we stand. There is an idea conveyed to the mind, of some universal and unlimited danger. We may flee from the crater of a volcano in active eruption, or from the dwelling whose destruction is threatened by the approach of the lava stream; but in an earthquake, direct our flight whithersoever we will, we still feel as if we trod upon the very focus of destruction." Not less striking is the testimony of Dr Tschudi, in his "Travels in Peru," regarding this singular effect of earthquakes on the human mind. "No familiarity with the phenomenon can," he remarks, "blunt the feeling. The inhabitant of Lima, who from childhood has frequently witnessed these convulsions of nature, is roused from his sleep by the shock, and rushes from his apartment with the cry of '*Misericordia!*' The foreigner from the north of Europe, who knows nothing of earthquakes but by description, waits with impatience to feel the movements of the earth, and longs to hear with his own ear the subterranean sounds, which he has hitherto considered fabulous. With levity he treats the apprehension of a coming convulsion, and laughs at the fears of the natives; but as soon as his wish is gratified, he is terror-striken, and is involuntarily prompted to seek safety in flight."

Now, a partially consolidated planet, tempested by frequent earthquakes of such terrible potency, that those of the historic ages would be but mere ripples of the earth's surface in comparison, could be no proper home for a creature so constituted. The fish or reptile,—animals of a limited range of instinct, exceedingly tenacious of life in most of their varieties, oviparous, prolific, and whose young immediately on their escape from the egg can provide for themselves, might enjoy existence in such circumstances, to the full extent of their narrow capacities; and when sudden death fell upon them,—though their remains, scattered over wide areas, continue to exhibit that distortion of posture incident to violent dissolution, which seems to speak of terror and suffering, —we may safely conclude there was but little real suffering in the case : they were happy up to a certain point, and unconscious for ever after. Fishes and reptiles were the proper inhabitants of our planet during the ages of the earth-tempests ; and when, under the operation of the chemical laws, these had become less frequent and terrible, the higher mammals were introduced. That prolonged ages of these tempests did exist, and that they gradually settled down, until the state of things became at length comparatively fixed and stable, few geologists will be disposed to deny. The evidence which supports *this* special theory of the development of our planet in its capabilities as a scene of organized and sentient being, seems palpable at every step. Look first at these Grauwacke rocks; and, after marking how in one place the strata have been upturned on their edges for miles together, and how in another the Plutonic rock has risen molten from below, pass on to the Old Red Sandstone, and examine its significant platforms of violent death,—its faults, displacements, and dislocations; see, next, in the Coal Measures, those evidences of sinking and ever-sinking strata, for thousands of feet together ; mark in the Oolite those vast overlying masses of trap, stretching athwart the landscape, far as the eye can reach ; observe carefully how the signs of convulsion and catastrophe gradually lessen as we descend to the times of the Tertiary, though even in these ages of the mammiferous quadruped, the earth must have had its oft-recurring ague fits of frightful intensity ; and then, on closing the survey, consider how exceedingly partial and unfrequent these earth-tempests have become in the recent periods. Yes ; we find everywhere marks of at once progression and identity,—of progress made, and yet identity maintained; but it is in the habitation that we find them,—not in the inhabitants. There is a tract of country in Hindustan that contains nearly as many square miles as all Great Britain, covered to the depth of hundreds of feet by one vast overflow of trap; a track similarly overflown, which exceeds in area all England, occurs in Southern Africa. The earth's surface is roughened with such,—mottled as thickly by the Plutonic masses as the skin of the leopard by its spots. The trap district which surrounds our Scottish metropolis, and imparts so imposing a character to its scenery, is too inconsiderable to be marked on geological maps of the world, that we yet see streaked and speckled with similar memorials, though on an immensely vaster scale, of the eruption and overflow which took place in the earthquake ages. What could man have done on the globe at a time when such outbursts were comparatively common occurrences? What could he have done where Edinburgh now stands during that overflow of trap porphyry of which the Pentland range forms but a fragment, or that outburst of greenstone of which but a portion remains in the dark ponderous coping of Salisbury Craigs, or when the thick floor of rock on which the city stands was broken up, like the ice of an arctic sea during a tempest in spring, and laid on edge from where it leans against the Castle Hill to beyond the quarries at Joppa? The reasoning brain would have been wholly at fault in a scene of things in which it could neither foresee the exterminating calamity while yet distant, nor control it when it had come; and so the reasoning brain was not produced until the scene had undergone a slow but thorough process of change, during which, at each progressive stage, it had furnished a platform for higher and still higher life.

When the coniferæ could flourish on the land, and fishes subsist in the seas, fishes and cone-bearing plants were created; when the earth became a fit habitat for reptiles and birds, reptiles and birds were produced; with the dawn of a more stable and mature state of things the sagacious quadruped was ushered in; and, last of all, when man's house was fully prepared for him,—when the data on which it is his nature to reason and calculate had become fixed and certain,—the reasoning, calculating brain was moulded by the creative finger, and man became a living soul. Such seems to be the true reading of the wondrous inscription chiselled deep in the rocks. It furnishes us with no clue by which to unravel the unapproachable mysteries of creation;—these mysteries belong to the wondrous Creator, and to Him only. We attempt to theorize upon them, and to reduce them to law, and all nature rises up against us in our presumptuous rebellion. A stray splinter of cone-bearing wood,—a fish's skull or tooth,—the vertebra of a reptile,—the humerus of a bird,—the jaw of a quadruped,—all, any of these things, weak and insignificant as they may seem, become in such a quarrel too strong for us and our theory: the puny fragment, in the grasp of truth, forms as irresistible a weapon as the dry bone did in that of Sampson of old; and our slaughtered sophisms lie piled up, "heaps upon heaps," before it.

There is no geological fact nor revealed doctrine with which this special scheme of development does not agree. To every truth, too, really such, from which the antagonist scheme derives its shadowy analogies, it leaves its full value. It has no quarrel with the facts of even the "Vestiges," in their character as realities. There is certainly something very extraordinary in that fœtal progress of the human brain on which the assertors of the development hypothesis have founded so much. Nature, in constructing this curious organ, first lays down a grooved cord, as the carpenter lays down the keel of his vessel; and on this narrow base the perfect brain, as month after month passes by, is gradually built up, like the vessel from the keel. First it grows up into a brain closely resembling that of a fish; a few additions more convert it into a brain undistinguishable from that of a reptile; a few additions more impart to it the perfect appearance of the brain of a bird; it then developes into a brain exceedingly like that of a mammiferous quadruped; and, finally, expanding atop, and spreading out its deeply corrugated lobes, till they project widely over the base, it assumes its unique character as a human brain. Radically such from the first, it passes towards its full development, through all the inferior forms, from that of the fish upwards,—thus comprising, during its fœtal progress, an epitome of geologic history, as if each man were in himself, not the *microcosm* of the old fanciful philosopher, but something greatly more wonderful,—a compendium of all animated nature, and of kin to every creature that lives. Hence the remark, that man is the sum total of all animals,—"the animal equivalent," says Oken, "to the whole animal kingdom." We are perhaps too much in the habit of setting aside real facts, when they have been first seized upon by the infidel, and appropriated to the purposes of unbelief, as if they had suffered contamination in his hands. We forget, like the brother "weak in the faith," instanced by the Apostle, that they are in themselves "creatures of God;" and too readily reject the lesson which they teach, simply because they have been offered in sacrifice to an idol. And this strange fact of the progress of the human brain is assuredly a fact none the less worth looking at from the circumstance that infidelity has looked at it first. On no principle recognisable in right reason can it be urged in support of the development hypothesis ;—it is a fact of *fœtal* development, and of that only. But it would be well should it lead our metaphysicians to enquire whether they have not been rendering their science too insulated and exclusive; and whether the mind that works by a brain thus "fearfully and wonderfully made," ought not to be viewed rather in connection with all animated nature, especially as we find nature exemplified in the various vertebral forms, than as a thing fundamentally abstract and distinct. The brain built up of all the types of *brain*, may be the organ of a mind compounded, if I may so express myself, of all the varieties of *mind*. It would be perhaps over fanciful to urge that it is the creature who has made himself free of all the elements, whose brain has been thus in succession that of all their proper denizens ; and that there is no animal instinct, the function of which cannot be illustrated by some art mastered by man : but there can be nothing over fanciful in the suggestion, founded on this fact of fœtal development, that possibly some of the more obscure signs impressed upon the human character may be best read through the spectacles of physical science. The successive phases of the fœtal brain give at least fair warning that, in tracing to its first principles the moral and intellectual nature of man, what is properly his "natural history" should not be overlooked. Oken, after describing the human creature in one passage as "equivalent to the whole animal kingdom," designates him in another as "God wholly manifested," and as "God become man ;"—a style of expression at which the English reader may start, as that of the "big mouth speaking blasphemy," but which has become exceedingly common among the rationalists of the Continent. The irreverent naturalist ought surely to have remembered, that the sum total of all the animals cannot be different in its nature from the various sums of which it is an aggregate,—seeing that *no* summation ever differs in *quality* from the items summed up, which compose it,—and that, though it may amount in this case to man *the animal*,—to man, as he may be weighed, and measured, and subjected to the dissecting knife,—it cannot possibly amount to God. Is God merely a sum total of birds and beasts, reptiles and fishes? —a mere Egyptian deity, composed of fantastic hieroglyphics derived from the forms of the brute creation? The impieties of the transcendentalist may, however, serve to illustrate that mode of seizing on terms which, as the most sacred in the message of revelation, have been long coupled in the popular mind with saving truths, and forcibly compelling them to bear some visionary and illusive meaning, wholly foreign to that with which they were originally invested, which has become so remarkable a part of the policy of modern infidelity. Rationalism has learned to sacrifice to Deity with a certain measure of conformity to the required pattern; but it is a conformity in appearance only, not in reality: the sacrifice always resembles that of Prometheus of old, who presented to Jupiter what, though it seemed to be an ox without blemish, was merely an ox-skin stuffed full of bones and garbage.

There is another very remarkable class of facts in geological history, which appear to fall as legitimately within the scope of argument founded on final causes, as those which bear on the appearance of man at his proper era. The period of the mammiferous quadrupeds seems, like the succeeding human period, to have been determined, as I have said, by the earth's fitness at the time as a place of habitation for creatures so formed. And the bulk to which, in the more extreme cases, they attained, appears to have been regulated, as in the higher mammals now, with reference to the force of gravity at the earth's surface. The Megatherium and the Mastodon, the Dinotherium and the extinct elephant, increased in bulk, in obedience to the laws of the specific constitution imparted to them at their creation ; and these laws bore reference, in turn, to another law,—that law of gravity which determines that no creature which moves in air and treads the surface of the earth should exceed a certain weight or size. To very near the limits assigned by this law some of the ancient quadrupeds arose. It is even doubtful whether the Dinotherium, the most gigantic of mammals, may not have been, like the existing sea-lions and morses, mainly an aquatic quadruped ;—an inference grounded on the circumstance that, in at least portions of its framework, it seems to have risen beyond these limits. Now, it does not seem wonderful that, with apparent reference to the point at which the gravity of bodies at the earth's surface *bisects* the conditions of texture and matter necessary to existence among the sub-aerial vertebrata, the *reptiles* of the Secondary periods should have grown up in some of their species and genera to the extreme size. A world of frogs, newts, and lizards would have borne stamped upon it the impress of a tame and miserable mediocrity, that would have harmonized ill with the extent of the earth's capabilities for supporting life on a large scale. There would be no principle of adaptation or rule of proportion maintained between an animal kingdom composed of so contemptible a group of beings, and either the dynamic laws under which matter exists on our planet, or the luxuriant vegetation which it bore during the Secondary ages. And such was not the character of the group which composed the reptile dynasty. The Iguanodon must have been quite as tall as the elephant,—greatly longer, and, it would seem, at least as bulky. The Megalosaurus must have at least equalled the rhinoceros; the Hylæosaurus would have outweighed the hippopotamus. And when reptiles that rivalled in size our hugest mammals inhabited the land, other reptiles,—Ichthyosaurs, Plesiosaurs, and Cetiosaurs,—scarce less bulky than the cetacea themselves, possessed the sea. Not only was the platform of being occupied in all its *breadth*, but also in all its *height;* and it is according to our simpler and more obvious ideas of adaptation,—simple and obvious because gleaned from the very surface of the universe of life,—that such should have been the case. But it does appear strange, because under the regulation, it would seem, of a principle of adaptation more occult, and, if I may so speak, more *Providential*, that no sooner are the huge mammals introduced *as a group*, than, with but a few exceptions, the reptiles appear in greatly diminished proportions. They no longer occupy the platform to its full extent of *height*. Even in tropical countries, in which certain families of mammals still attain to the maximum size, the reptiles, if we except the crocodilean family, a few harmless turtles, and the degraded boas and pythons, are a small and comparatively unimportant race. Nay, the existing giants of the class,—the crocodiles and boas,—hardly equal in bulk the third-rate reptiles of the ages of the Oolite and the Wealden. So far as can be seen, there is no reason deduceable from the nature of things, why the country that sustains a mammal bulky as the elephant, should not also support a reptile huge as the Iguanodon; or why the Megalosaurus, Hylæosaurus, and Dicynodon, might not have been contemporary with the lion, tiger, and rhinoceros. The change which took place in the reptile group immediately on their dethronement at the close of the Secondary period, seems scarce less strange than that sung by Milton :

> "Behold a wonder ! They but now who seemed
> In bigness to surpass earth's giant sons,
> Now less than smallest dwarfs, in narrow room
> Thronged numberless ; like that pygmean race
> Beyond the Indian mount ; or fairy elves,
> Whose midnight revels, by a forest side
> Or fountain, some belated peasant sees,
> Or dreams he sees, while, overhead, the moon
> Sits arbitress, and nearer to the earth
> Wheels her pale course."

But though we cannot assign a *cause* for this general reduction of the reptile class, save simply the will of the all-wise Creator, the *reason* why it should have taken place seems easily assignable. It was a bold saying of the old philosophic heathen, that "God is the soul of brutes ;" but writers on instinct in even our own times have said less warrantable things. God *does* seem to do for many of the inferior animals of the lower divisions, that, though devoid of brain and vertebral column, are yet skilful chemists and accomplished architects and mathematicians, what he enables man, through the exercise of the reasoning faculty, to do for himself ; and the ancient philosopher meant no more. And in clearing away the giants of the reptile dynasty, when their kingdom had passed away, and then re-introducing the class as much shrunken in their proportions as restricted in their domains, the Creator seems to have been doing for the mammals what man, in the character of a "mighty hunter before the Lord," does for himself. There is in nature very little of what can be called war. The cities of this country cannot be said to be in a state of war, though their cattle-markets are thronged every week with animals for slaughter, and the butcher and fish-monger find their places of business thronged with customers. And such, in the main, is the condition of the animal world ;—it consists of its two classes,—animals of prey, and the animals upon which they prey : its wars are simply those of the butcher and fisher, lightened by a dash of the enjoyments of the sportsman.

> "The creatures see of flood and field,
> And those that travel on the wind,
> With them no strife can last; they live
> In peace and peace of mind."

Generally speaking, the carnivorous mammalia respect one another : lion does not war with tiger, nor the leopard contend with the hyena. But the carnivorous reptiles manifest no such respect for the carnivorous mammals. There are fierce contests in their native jungles, on the banks of the Ganges, between the gavial and the tiger ; and in the steaming forests of South America, the boa-constrictor casts his terrible coil scarce less readily round the puma than the antelope. A world which, after it had become a home of the higher herbivorous and more powerful carnivorous mammals, continued to retain the gigantic reptiles of its earlier ages, would be a world of horrid, exterminating war, and altogether rather a place of torment than a scene of intermediate character, in which, though it sometimes re-echoes the groans of suffering nature, life is, in the main, enjoyment. And so,—save in a few exceptional cases, that, while they establish the rule as a fact, serve also as a key to unlock that principle of the Divine government on which it appears to rest,—no sooner was the reptile removed from his place in the fore-front of creation, and creatures of a higher order introduced into the consolidating and fast-ripening planet of which he had been so long the monarch, than his bulk shrank and his strength lessened, and he assumed a humility of form and aspect at once in keeping with his reduced circumstances, and compatible with the general welfare. But though the *reason* of the reduction appears obvious, I know not that it can be referred to any other *cause* than simply the will of the All-Wise Creator.

There hangs a mystery greatly more profound over the fact of the *degradation* than over that of the *reduction* and *diminution* of classes. We can assign what at least *seems* to be a sufficient *reason* why, when reptiles formed as a class the highest representatives of the vertebrata, they should be of imposing bulk and strength, and altogether worthy of that post of precedence which they then occupied among the animals. We can also assign a *reason* for the strange reduction which took place among them in strength and bulk immediately on their removal from the first to the second place. But why not only *reduction*, but also *degradation?* Why, as division started up in advance of division,—first the reptiles in front of the fishes, then the quadrupedal mammals in front of the reptiles, and, last of all, man in front of the quadrupedal mammals,—should the supplanted classes,—two of them at least—fishes and reptiles—for there seem to have been no additions made to the mammals since man entered upon the scene,—why should they have become the receptacles of orders and families of a degraded character, which had no place among them in their monarchical state? The fishes removed beyond all analogy with the higher vertebrata, by their homocercal tails,—the fishes (*Acanthopterygii* and *Sub-brachiati*) with their four limbs slung in a belt round their necks,—the flat fishes (*Pleuronectidæ*), that, in addition to this deformity, are so twisted to a side, that while the one eye occupies a single orbit in the middle of the skull, the other is thrust out to its edge,—the irregular fishes generally (sun-fishes, frog-fishes, hippocampi, &c.) were not introduced into the ichthyic division until after the full development of the reptile dynasty; nor did the hand that makes no slips in its working "form the crooked serpent," footless, grovelling, venom-bearing,—the authorized type of a fallen and degraded creature,—until after the introduction of the mammals. What can this fact of degradation mean? Species and genera seem to be greatly more numerous in the present age of the world than in any of the geologic ages. Is it not possible that the extension of the chain of being which has thus taken place,—not only, as we find, through the addition of the higher divisions of animals to its upper end, but also through the interpolations of *lower links* into the previously existing divisions,—may have borne reference to some predetermined scheme of well-proportioned gradation, or, according to the poet,

"Of general ORDER since the whole began?"

May not, in short, what we term degradation be merely one of the modes resorted to for filling up the voids in creation, and thereby perfecting a scale which must have been originally not merely a scale of narrow compass, but also of innumerable breaks and blanks, hiatuses and chasms? Such, certainly, would be the reading of the enigma which a Soame Jenyns or a Bolingbroke would suggest; but the geologist has learned from his science, that the completion of a chain of at least contemporary being, perfect in its gradations, cannot possibly have formed the design of Providence. Almost ever since God united vitality to matter, the links in this chain of animated nature, as if composed of a material too brittle to bear their own weight when stretched across the geologic ages, have been dropping one after one from his hand, and sinking, fractured and broken, into the rocks below. It is urged by Pope, that were "we to press on superior powers," and rise from our own assigned place to the place immediately above it, we would, in consequence of the transposition,

> "In the full creation leave a void,
> Where, one step broken, the great scale's destroyed.
> From nature's chain whatever link we strike,
> Tenth or ten thousandth, breaks the chain alike."

The poet could scarce have anticipated that there was a science then sleeping in its cradle, and dreaming the dreams of Whiston, Leibnitz, and Burnet, which was one day to rise and demonstrate that both the tenth and the ten thousandth link in the chain had been already broken and laid by, with all the thousands of links between ; and that man might laudably "press on superior powers," and attain to a "new nature," without in the least affecting the symmetry of creation by the void which his elevation would necessarily create ; that, in fine, voids and blanks in the scale are exceedingly common things ; and that if men could, by rising into angels, make one blank more, they might do so with perfect impunity. Farther, even were the graduated chain of Bolingbroke a reality, and not what Johnson well designates it, an "absurd hypothesis," and were what I have termed the interpolation of links necessary to its completion, the mere filling up of the original blanks and chasms would not necessarily involve the fact of degradation, seeing that each blank could be filled up, if I may so express myself, from its lower end. Each could be as certainly occupied to the full by an elevation of lower forms, as by a humiliation of the higher. We might receive the hypothesis of Bolingbroke, and yet find the mysterious fact of degradation remain an unsolved riddle in our hands.

But though I can assign neither *reason* nor *cause* for the fact, I cannot avoid the conclusion, that it is associated with certain other great facts in the moral government of the universe, by those threads of analogical connection which run through the entire tissue of Creation and Providence, and impart to it that character of unity which speaks of the single producing Mind. The first idea of every religion on earth which has arisen out of what may be termed the spiritual instincts of man's nature, is that of a Future State ; the second idea is, that in this state men shall exist in two separate classes, —the one in advance of their present condition, the other far in the rear of it. It is on these two great beliefs that conscience everywhere finds the fulcrum from which it acts upon the conduct ; and it is, we find, wholly inoperative as a force without them. And in that one religion among men that, instead of retiring, like the pale ghosts of the others, before the light of civilization, brightens and expands in its beams, and in favour of whose claim as a revelation from God the highest philosophy has declared, we find these two master ideas occupying a still more prominent place than in any of those merely indigenous religions that spring up in the human mind of themselves. The special lesson which the Adorable Saviour, during his ministry on earth, oftenest enforced, and to which all the others bore reference, was the lesson of a final separation of mankind into two great divisions,—a division of God-like men, of whose high standing and full-orbed happiness man, in the present scene of things, can form no adequate conception ; and a division of men finally lost, and doomed to unutterable misery and hopeless degradation.

There is not in all Revelation a single doctrine which we find oftener or more clearly enforced than that there shall continue to exist, throughout the endless cycles of the future, a race of degraded men and of degraded angels.

Now, it is truly wonderful how thoroughly, in its general scope, the revealed pieces on to the geologic record. We know, as geologists, that the dynasty of the fish was succeeded by that of the reptile,—that the dynasty of the reptile was succeeded by that of the mammiferous quadruped,—and that the dynasty of the mammiferous quadruped was succeeded by that of man as man now exists,—a creature of mixed character, and subject, in all conditions, to wide alternations of enjoyment and suffering. We know, further,—so far at least as we have yet succeeded in deciphering the record,—that the several dynasties were introduced, not in their lower, but in their higher forms;—that, in short, in the imposing programme of creation it was arranged, as a general rule, that in each of the great divisions of the procession the magnates should walk first. We recognise yet further the fact of degradation specially exemplified in the fish and the reptile. And then, passing on to the revealed record, we learn that the dynasty of man in the mixed state and character is not the final one, but that there is to be yet another creation, or, more properly, *re*-creation, known theologically as the Resurrection, which shall be connected in its physical components, by bonds of mysterious paternity, with the dynasty which now reigns, and be bound to it mentally by the chain of identity, conscious and actual; but which, in all that constitutes superiority, shall be as vastly its superior as the dynasty of responsible man is superior to even the lowest of the preliminary dynasties. We are further taught, that at the commencement of this last of the dynasties there will be a re-creation of not only elevated, but also of degraded beings,—a re-creation of the *lost*. We are taught yet further, that though the present dynasty be that of a lapsed race, which at their first introduction were placed on higher ground than that on which they now stand, and sank by their own act, it was yet part of the original design, from the beginning of all things, that they should occupy the existing platform; and that Redemption is thus no after-thought, rendered necessary by the Fall, but, on the contrary, part of a general scheme, for which provision had been made from the beginning; so that the Divine Man, through whom the work of restoration has been effected, was in reality, in reference to the purposes of the Eternal, what he is designated in the remarkable text, "*the Lamb slain from the foundations of the world.*" Slain from the foundations of the world! Could the assertors of the stony science ask for language more express? By piecing the two records together,—that revealed in Scripture and that revealed in the rocks,—records which, however widely geologists may mistake the one, or commentators misunderstand the other, have emanated from the same great Author,—we learn that in slow and solemn majesty has period succeeded period, each in succession ushering in a higher and yet higher scene of existence,—that fish, reptiles, mammiferous quadrupeds, have reigned in turn,—that responsible man, "made in the image of God," and with dominion over all creatures, ultimately entered into a world ripened for his reception; but, further, that this passing scene, in which he forms the prominent figure, is not the final one in the long series, but merely the last of the *preliminary* scenes; and that that period to which the bygone ages, incalculable in amount, with all their well-proportioned gradations of being, form the imposing vestibule, shall have perfection for its occupant, and eternity for its duration. I know not how it may appear to others; but for my own part, I cannot avoid thinking that there would be a lack of proportion in the series of being, were the period of perfect and glorified humanity abruptly connected, without the introduction of an intermediate creation of *responsible* imperfection, with that of the dying irresponsible brute. That scene of things in which God became Man, and suffered, *seems*, as it no doubt *is*, a necessary link in the chain.

I am aware that I stand on the confines of a mystery which man, since the first introduction of sin into the world till now, has "vainly aspired to comprehend." But I have no new reading of the enigma to offer. I know not why it is that moral evil exists in the universe of the All-Wise and the All-Powerful; nor through what occult law of Deity it is that "perfection should come through suffering." The question, like that satellite, ever attendant upon our planet, which presents both its sides to the sun, but invariably the same side to the earth, hides one of its faces from man, and turns it to but the Eye from which all light emanates. And it is in that God-ward phase of the question that the mystery dwells. We can map and measure every protuberance and hollow which roughens the nether disk of the moon, as, during the shades of night, it looks down upon our path to cheer and enlighten; but what can we know of the other? It would, however, seem, that even in this field of mystery the extent of the inexplicable and the unknown is capable of reduction, and that the human understanding is vested in an ability of progressing towards the central point of that dark field throughout all time, mayhap all eternity, as the asymptote progresses upon its curve. Even though the essence of the question should for ever remain a mystery, it may yet, in its reduced and defined state, serve as a key for the laying of other mysteries open. The philosophers are still as ignorant as ever respecting the intrinsic nature of gravitation; but regarded simply as a force, how many enigmas has it not served to unlock! And that moral gravitation towards evil, manifested by the only two classes of responsible beings of which there is aught known to man, and of which a degradation linked by mysterious analogy with a class of facts singularly prominent in geologic history is the result, occupies apparently a similar place, as a force, in the moral dynamics of the universe, and seems suited to perform a similar part. Inexplicable itself, it is yet a key to the solution of all the minor inexplicabilities in the scheme of Providence.

In a matter of such extreme niceness and difficulty, shall I dare venture on an illustrative example?

So far as both the geologic and the Scriptural evidence extends, no species or family of existences seems to have been introduced by creation into the present scene of being since the appearance of man. In Scripture the formation of the human race is described as the terminal act of a series, "good" in all its previous stages, but which became "very good" then; and geologists, judging from the modicum of evidence which they have hitherto succeeded in collecting on the subject,—evidence still meagre, but, so far as it goes, independent and distinct,—pronounce "post-Adamic creations" at least "improbable." The naturalist finds certain animal and vegetable species restricted to certain circles, and that in certain foci in these circles they attain to their fullest development and their maximum number. And these foci he regards as the original centres of creation, whence, in each instance in the process of increase and multiplication, the plant or creature propagated itself outwards in circular wavelets of life, that sank at each stage as they widened, till at length, at the circumference of the area, they wholly ceased. Now we find it argued by Professor Edward Forbes, that "since man's appearance, certain geological areas, both of land and water, have been formed, presenting such physical conditions as to entitle us to expect within their bounds one, or in some instances more than one, centre of creation, or *point of maximum of a zoological or botanical province.* But a critical examination renders evident," the Professor adds, "that instead of showing distinct foci of creation, they have been in all instances peopled by colonization, *i. e.* by migration of species from pre-existing, and in every case pre-Adamic, provinces. Among the terrestrial areas the British isles may serve as an example; among marine, the Baltic, Mediterranean, and Black Seas. The British islands have been colonized from various centres of creation in (now) continental Europe; the Baltic Sea from the Celtic region, although it runs itself into the conditions of the Boreal one; and the Mediterranean, as it now appears, from the fauna and flora of the more ancient Lusitanian province." Professor Forbes, it is stated farther, in the report of his paper to which I owe these details,—a paper read at the Royal Institution in March last,—"exhibited, in support of the same view, a map, showing the relation which the centres of creation of the air-breathing molluscs in Europe bear to the geological history of the respective areas, and proving that the whole snail population of its northern and central extent (the portion of the Continent of newest and probably post-Adamic origin) had been derived from foci of creation seated in pre-Adamic lands. And these remarkable facts have induced the Professor," it was added, "to maintain the improbability of post-Adamic creations."

With the introduction of man into the scene of existence, creation, I repeat, seems to have ceased. What is it that now takes its place, and performs its work? During the previous dynasties, all elevation in the scale was an effect simply of creation. Nature lay dead in a waste theatre of rock, vapour, and sea, in which the insensate laws, chemical, mechanical, and electric, carried on their blind, unintelligent processes: the *creative fiat* went forth; and, amid waters that straightway teemed with life in its lower forms, vegetable and animal, the dynasty of the fish was introduced. Many ages passed, during which there took place no farther elevation: on the contrary, in not a few of the newly introduced species of the reigning class there occurred for the first time examples of an asymmetrical misplacement of parts, and, in at least one family of fishes, instances of defect of parts: there was the manifestation of a downward tendency towards the degradation of monstrosity, when the elevatory fiat again went forth, and, *through an act of creation,* the dynasty of the reptile began. Again many ages passed by, marked, apparently, by the introduction of a warm-blooded oviparous animal, the bird, and of a few marsupial quadrupeds, but in which the prevailing class reigned undeposed, though at least unelevated. Yet again, however, the elevatory fiat went forth, and *through an act of creation* the dynasty of the mammiferous quadruped began. And after the further lapse of ages, the elevatory fiat went forth yet once more *in an act of creation;* and with the human, heaven-aspiring dynasty, the moral government of God, in its connection with at least the world which we inhabit, "took beginning." And then creation ceased. Why? Simply because God's moral government *had* begun,—because in necessary conformity with the institution of that government, there was to be a thorough identity maintained between the glorified and immortal beings of the terminal dynasty, and the dying magnates of the dynasty which now is; and because, in consequence of the maintenance of this identity as an essential condition of this moral government, mere *acts of creation* could no longer carry on the elevatory process. The work analogous in its end and object to those *acts of creation* which gave to our planet its successive dynasties of higher and yet higher existences, is the work of REDEMPTION. It is the elevatory process of the present time,—the only possible provision for that final act of *re*-creation "to everlasting life," which shall usher in the terminal dynasty.

I cannot avoid thinking that many of our theologians attach a too narrow meaning to the remarkable reason "annexed to the Fourth Commandment" by the Divine Lawgiver. "God rested on the seventh day," says the text, "from all his work which He had created and made; and God blessed the seventh day, and sanctified it." And such is the reason given in the Decalogue why man should also rest on the seventh day. God rested on the Sabbath, and sanctified it; and therefore man ought also to rest on the Sabbath, and keep it holy. But I know not where we shall find grounds for the belief that that Sabbath-day during which God rested was merely commensurate in its duration with one of the Sabbaths of short-lived man,—a brief period, measured by a single revolution of the earth on its axis. We have not, as has been shown, a shadow of evidence that He resumed his work of creation on the morrow: the geologist finds no trace of post-Adamic creation,—the theologian can tell us of none. God's Sabbath of rest may still exist;—*the work of* REDEMPTION *may be the work of his Sabbath day.* That elevatory process through successive acts of creation which engaged Him during myriads of ages, was of an ordinary week-day character ; but when the term of his moral government began, the elevatory process proper to it assumed the Divine character of the Sabbath. This special view appears to lend peculiar emphasis to the reason embodied in the commandment. The collation of the passage with the geologic record seems, as if by a species of re-translation, to make it enunciate as its injunction, "Keep this day, not merely as a day of memorial related to a past fact, but also as a day of co-operation with God in the work of elevation in relation both to a present fact and a future purpose. God keeps his Sabbath," it says, "in order that He may save ; keep yours also, in order that ye may be saved." It serves, besides, to throw light on the prominence of the Sabbatical command, in a digest of law of which no part or tittle can pass away until the fulfilment of all things. During the present dynasty of probation and trial, that special work of both God and man on which the character of the future dynasty depends, is the Sabbath-day work of saving and being saved.*

---

* The common objection to that special view which regards the *days* of creation as immensely protracted periods of time, furnishes a specimen, if not of reasoning in a circle, at least of reasoning from a mere assumption. It first takes for granted, that the Sabbath-day during which God rested was a day of but twenty-four hours; and then argues, from the supposition, that in order to *keep up the proportion* between the six previous working days and the seventh day of rest, which the reason annexed to the fourth commandment demands, these previous days must *also* have been days of twenty-four hours each. It would, I have begun to suspect, square better with the ascertained facts, and be at least equally in accordance with Scripture, to reverse the process, and argue that, *because* God's working days

It is in this dynasty of the future that man's moral and intellectual faculties will receive their full development. The expectation of any very great advance in the present

---

were immensely protracted periods, *his* Sabbath must *also* be an immensely protracted period. The reason attached to the law of the Sabbath seems to be simply *a reason of proportion*;—the objection to which I refer is an objection palpably founded on *considerations* of proportion. And certainly, were the reason to be divested of proportion, it would be divested also of its distinctive character as a reason. Were it to run as follows, it could not be at all understood:—"Six days shalt thou labour, &c., but on the seventh day shalt thou do no labour, &c.; for in six immensely protracted periods of many thousand years each did the Lord make the heavens and earth, &c., and then rested during a brief day of twenty-four hours; therefore the Lord blessed the brief day of twenty-four hours, and hallowed it." This, I repeat, would not be reason. All, however, that seems necessary to the integrity of the reason, in its character as such, is, that the proportion of six parts to seven should be maintained. God's periods may be periods expressed algebraically by letters symbolical of unknown quantity, and man's periods by letters symbolical of quantities well known; but if God's Sabbath be equal to one of his six working days, and man's Sabbath equal to one of *his* six working days, the integrity of proportion is maintained. When I see the palpable absurdity of such a reading of the reason as the one given above, I can see no absurdity whatever in the reading which I subjoin:—"Six *periods* $(a=a=a=a=a=a)$ shalt thou labour, &c., but on the seventh *period* $(b=a)$ shalt thou do no labour, &c.; for in six *periods* $(x=x=x=x=x=x)$ the Lord made heaven and earth, &c., and rested the seventh *period* $(y=x)$; therefore the Lord blessed the seventh *period*, and hallowed it. The reason, in its character as a reason of proportion, survives here in all its integrity. Man, when in his unfallen estate, bore the image of God, but it must have been a miniature image at best;—the proportion of man's week to that of his Maker may, for aught that appears, be mathematically just in its proportions, and yet be a miniature image too,—the mere scale of a map, on which inches represent geographical degrees. All those week-days and Sabbath-days of man which have come and gone since man first entered

scene of things,—great, at least, when measured by man's large capacity of conceiving of the good and fair,—seems to be, like all human hope when restricted to time, an expectation doomed to disappointment. There are certain limits within which the race improves;—civilization is better than the want of it, and the taught superior to the untaught man. There is a change, too, effected in the moral nature, through that Spirit which, by working belief in the heart, brings its aspirations into harmony with the realities of the unseen world, that, in at least its relation to the future state, cannot be estimated too highly. But conception can travel very far beyond even its best effects in their merely secular bearing; nay, it is peculiarly its nature to show the men most truly the subjects of it, how miserably they fall short of the high standard of conduct and feeling which it erects, and to teach them, more emphatically than by words, that their degree of happiness must of necessity be as low as their moral attainments are humble. Further,—man, though he has been increasing in knowledge ever since his appearance on earth, has not been improving in faculty;—a shrewd fact, which they who expect most from the future of this world would do well to consider. The ancient masters of mind were in no respect inferior in calibre to their predecessors. We have not yet shot ahead of the old Greeks in either the perception of the beautiful, or in the ability of producing it; there has been no improvement in the inventive faculty since the Iliad was written, some three thousand years ago; nor has taste become

---

upon this scene of being, with all which shall yet come and go, until the resurrection of the dead terminates the work of Redemption, may be included, and probably *are* included, in the one Sabbath-day of God.

more exquisite, or the perception of the harmony of numbers more nice, since the age of the Æneid. Science is cumulative in its character ; and so its votaries in modern times stand on a higher pedestal than their predecessors. But though nature produced a Newton some two centuries ago, as she produced a Goliath of Gath at an earlier period, the modern philosophers, as a class, do not exceed in actual stature the worse informed ancients,—the Euclids, Archimedeses, and Aristotles. We would be without excuse if, with the Bacon, Milton, and Shakspeare of these latter ages of the world full before us, we recurred to the obsolete belief that the human race is deteriorating ; but then, on the other hand, we have certain evidence, that since genius first began unconsciously to register in its works its own bulk and proportions, there has been no increase in the mass or improvement in the quality of individual mind. As for the dream that there is to be some extraordinary elevation of the general platform of the race achieved by means of education, it is simply the hallucination of the age,—the world's present alchemical expedient for converting farthings into guineas, sheerly by dint of scouring. Not but that education is good : it exercises, and, in the ordinary mind, developes, faculty. But it will not anticipate the terminal dynasty. Yet further,—man's average capacity of happiness seems to be as limited and as incapable of increase as his average reach of intellect : it is a mediocre capacity at best ; nor is it greater by a shade now, in these days of power-looms and portable manures, than in the times of the old patriarchs. So long, too, as the law of increase continues, man must be subject to the law of death, with its stern attendants, suffering and sorrow ; for the two laws go necessarily together ; and so long as death reigns, human creatures, in even the best of times, will continue to quit this scene of being without professing much satisfaction at what they have found either in it or themselves. It will no doubt be a less miserable world than it is now, when the good come, as there is reason to hope they one day shall, to be a majority ; but it will be felt to be an inferior sort of world even then, and be even fuller than now of wishes and longings for a better. Let it improve as it may, it will be a scene of probation and trial till the end. And so Faith, undeceived by the mirage of the midway desert, whatever form or name, political or religious, the phantasmagoria may bear, must continue to look beyond its unsolid and tremulous glitter,—its bare rocks exaggerated by the vapour into air-drawn castles, and its stunted bushes magnified into goodly trees,—and, fixing her gaze upon the re-creation yet future,—the terminal dynasty yet unbegun, —she must be content to enter upon her final rest,—for she will not enter upon it earlier,—" at return"

" Of Him, the Woman's Seed,
Last in the clouds, from heaven to be revealed
In glory of the Father, to dissolve
Satan with his perverted world, then raise
From the conflagrant mass, purged and refined,
New heavens, new earth, ages of endless date,
Founded in righteousness, and peace, and love,
To bring forth fruits,—joy and eternal bliss."

But it may be judged that I am trespassing on a field into which I have no right to enter. Save, however, for its close proximity with that in which the geologist expatiates as properly his own, this little volume would never have been written. It is the fact that man must believingly co-operate with God in the work of preparation for the final dynasty, or exist throughout its never-ending cycles as a lost and degraded creature, that alone renders the development hypothesis formidable. By inculcating that the elevatory process is one of natural law, not of moral endeavour,—by teaching, inferentially at least, that in the better state of things which is coming there is to be an identity of race with that of the existing dynasty, but no identity of individual consciousness,—that, on the contrary, the life after death which we are to inherit is to be merely a horrid life of wriggling impurities, originated in the putrefactive mucus,—and that thus the men who now live possess no real stake in the kingdom of the future, —it is its direct tendency, so far as its influence extends, to render the required co-operation with God an impossibility. For that co-operation cannot exist without belief as its basis. The hypothesis involves a misreading of the geologic record, which not merely affects its meaning in relation to the mind, and thus, in a question of science, substitutes error for truth, but which also threatens to affect the record itself, in relation to the destiny of every individual perverted and led astray. It threatens to write down among the degraded and the lost, men who, under the influence of an unshaken faith, might have risen at the dawn of the terminal period, to enjoy the fulness of eternity among the glorified and the good.

THE END.

Printed in the United States
By Bookmasters